# LEONARDO'S FOOT

## HOW 10 TOES, 52 BONES, AND 66 MUSCLES

### SHAPED THE HUMAN WORLD

# LEONARDO'S FOOT

## HOW 10 TOES, 52 BONES, AND 66 MUSCLES
### SHAPED THE HUMAN WORLD

CAROL ANN RINZLER

BELLEVUE LITERARY PRESS
New York

First published in the United States in 2013 by
Bellevue Literary Press, New York

FOR INFORMATION CONTACT:
Bellevue Literary Press
NYU School of Medicine
550 First Avenue
OBV A612
New York, NY 10016

Bellevue Literary Press would like to thank all its generous
donors—individuals and foundations—for their support.

Library of Congress Cataloging-in-Publication Data
Rinzler, Carol Ann.
Leonardo's foot : how 10 toes, 52 bones, and 66 muscles shaped the human
world / Carol Ann Rinzler. — 1st ed.
p. ; cm.
Includes bibliographical references and index.
ISBN 978-1-934137-62-8 (pbk.)
I. Title.
[DNLM: 1. Anthropology, Physical—history—Popular Works. 2. Biological
Evolution--Popular Works. 3. Foot—Popular Works. 4. History of Medicine—
Popular Works. GN 50.4]
QP34.5
612—dc23
2013004519

Book design and type formatting by Mulberry Tree Press, Inc.
Manufactured in the United States of America

First Edition

10 9 8 7 6 5 4 3 2 1

ISBN 978-1-934137-62-8 pb

*"The human foot is a masterpiece of engineering and a work of art."*

Leonardo da Vinci, *The Notebooks* (c. 1508–1518)

*For*

Perry Luntz

MARCH 9, 1927–APRIL 13, 2009

*and*

Patricia Marie Dolan

MAY 6, 1939–NOVEMBER 15, 2011

# CONTENTS

# ILLUSTRATIONS

# LEONARDO'S FOOT

## HOW 10 TOES, 52 BONES, AND 66 MUSCLES

### SHAPED THE HUMAN WORLD

# INTRODUCTION

BოოKS HAPPEN.

You read something or a friend says something or you're walking down the street and you see something and you say to yourself, that's *interesting*. Then everywhere you look you see something about the *something*, and sooner or later the *something* turns into an idea and the idea turns into a proposal and then, if you're lucky, into a book.

One August morning in 2011, as I was lacing up my sneakers, I looked down at my underwhelming, underreported, and completely indispensable human feet, and thought, "That's *interesting*."

When I told my agent, Phyllis Westberg, that I had decided to write about feet, her response was not what you would call encouraging. Phyllis and I have lived through more than twenty books together, so I take seriously her view of authoring which is that if you can't put it on paper, it isn't real. And when she says paper, that's what she means—computer screens don't count. So I started to put it on paper, and four months later, I had two chapters and Phyllis said, "Who knew feet could be this interesting."

Actually, I did.

And so, it turned out, did the authors of the Old and New Testaments, Leonardo da Vinci, Ben Franklin, Sigmund Freud, and virtually every single twentieth century anthropologist who wandered through Africa, Asia, and Europe in search of the first primate to stand up on two legs.

The only body part completely exclusive to human beings is the chin (more about that later on). Everything else—eyes, ears, nose, heart, lungs, liver, arms, and legs—can be found elsewhere in the animal kingdom. That includes our hand with its famous opposable thumb. Apes, giant pandas, raccoons, and opossums also have an opposable thumb; koalas have not one, but two opposable thumbs on their front paws (the back paws have only one).

1

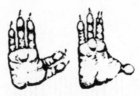

**Koala hand with two thumbs (*left*) vs. foot with one thumb (*right*).**

But like the chin, the homely foot with its adducted big toe, firm arch, and plantigrade sole, different in design from every other foot and hoof on earth, is unique and much to our advantage, powering our movement and, throughout history, enabling our cultural, political, and scientific development.

This was not always a popular idea. As my editor Leslie Hodgkins observes, our insistence on the evolutionary primacy of the brain said (and says) much about how we humans see ourselves. Our brain seems to us the thing that separates us from the herd on Noah's Ark while our body ties us to the rest of the world's inhabitants, and our foot, lowest of the low, binds us to the earth, the ground cursed with thorns and thistles because Eve ate that apple. Yes, we speculate that this or that animal has an *almost-human* brain. But we don't believe it for a minute, just as we don't believe that intelligence is so random that if—as the classic anecdote proposes—we put a million monkeys in a room with a million typewriters and leave them there long enough one of them will eventually type out the complete works of Shakespeare.

I write about food and health and medicine, which means I usually work surrounded by stacks of medical books, journals, and papers held in place by my trusty *Harrison's Principles of Internal Medicine*, one of the world's most erudite paperweights.

This time there were also history books to tell me how a clubfoot was one birth defect not considered justification for infanticide. There were art books with pictures of paintings and sculpture to show our fascination with a perfectly arched foot. There were nutrition texts to explain the relationship between protein, purines, and gout, the world's worst pain in the toe. There was psychiatry (and carefully selected pornography) to illustrate the sexual power of the fragrant foot.

There were biographical dictionaries to track a cast of characters ranging from Greek philosophers to Arab physicians, British poets, and American statesmen, all of whom had more than a word to say about our lower extremity. Where they are available—not all of them are—I have tacked a birth and death date onto the name of every important historical player in this story (living persons are entitled to their privacy) because knowing when they lived adds context to their actions. On occasion, I have included middle names. "Charles Robert Darwin" just sounds so much more friendly and accessible than plain "Charles Darwin." And you are free to make what you will of the fact that the great French essayist Michel Eyquem de Montaigne dropped what looks like a middle name, but is actually his father's family name, preferring to be known instead as Michel de Montaigne, Michael of the Chateau Montaigne, the place where he was born and the property he inherited in the town now known as *Saint-Michel-de-Montaigne*.

There were also two Bibles replete with references to the foot, most of them euphemisms for other body parts; *Bartlett's Familiar Quotations* as back-up; and the infinite Internet that grants access to everything from the staid old/new *Encyclopaedia Britannica* to a map of the town in Hanover County, Virginia, U.S.A., once nastily named with the detestable N-word.

Finally, there was my own particular pleasure, etymology. The linear ancestor of English is Old- or Anglo-Saxon English that, like Dutch and the Scandinavian languages, is an offshoot of German. But our current American English dictionary also includes words bequeathed to us by other languages, especially our scientific vocabulary, much of which dates to a time when Latin was the language of higher learning across the Western world. Throughout this book, when I use one of these words, I have attached its derivation, which sometimes comes as a surprise. Did you know, for example, that *Amazon,* the name for those ladies purported to have sliced off one breast so as to be able to draw a bow more efficiently across the chest, descends not from the relatively familiar Greek *a-* meaning *without* and *mazos* meaning *breast,* but from the much less well-known Persian *ha-maz-an* that translates roughly as *fighting together as one?*

Finding such gems is exciting. Wanting to share them is why writers write. Being able to do so in a way that has real, not virtual, weight, is why we write books.

In 1897, when James Ross Clemens, cousin of Samuel Langhorne Clemens, a.k.a. Mark Twain, died, reporters confused the two and ended up at the door of the more famous Clemens in search of a story only to be told, "The reports of my death are greatly exaggerated." You could say the same thing about printed books, the real ones with paper pages. Those of us who have made them our working life are used to hearing about their demise. In fact, we often swear the first such report surfaced about 15 minutes after Johann Gutenberg ran the first book, a Latin-language Bible, off his newly invented movable-type press and put it on sale in March 1455.

But we book people are a stubborn lot. Thanks to publishers like Erika Goldman of the Bellevue Literary Press, who maintain a serious, old-fashioned (in a good way) attention to books, and editors like Leslie Hodgkins, who know before we do where we want our stories to go, we and our books are still here.

As every publisher, and editor, and writer knows, we do not build our books alone.

In this case, LaRay Brown of the New York City Health and Hospitals Corporation; Thomas Blanck, Abraham Chachoua, Kathryn Coichetti-Mazzeo, Doris Farrelly, Susan Firestone, Kimberly Glassman, Maureen FitzPatrick, Leora Lowenthal, Robert Press, Raymonda Rastegar, Mel Rosenfeld, Nina Setia, Sandro Sherrod, and Maxine Simon at NYU Langone Medical Center; and Alex Bekker of the University of Medicine and Dentistry of New Jersey-Medical School offered valuable insight into the work of three extraordinary institutions. Rafael Tamargo of Johns Hopkins Hospital and Duncan MacRae, managing editor of *Neurosurgery*, eased the path to permission to reprint the wonderful pictures of Michelangelo's painting of the human brain in the head of God on the ceiling of the Sistine Chapel (yes, Michelangelo, like Leonardo, had something important to say—show—about the human foot). A. Barton Hinkle of the *Richmond Times-Dispatch* added details to the story of flat feet and southern town names. Jerry A. Cohen and Scott Groudine of the American Society of Anesthesiologists guided me through a project I followed while writing this book; one day, that, too, may build a book.

Minna Elias walked me through the world of godly feet, ancient and modern. Peter Sass read the material linking our feet and our psyche; Maria DeVal, for clubfoot and Lord Byron. Ellen Imbimbo gave her cool appraisal of philosophy as it sometimes relates to the foot. Louise Dankberg, Carol Greitzer, Linda Hoffman, Barbara Kloberdanz, and Trudy Mason were willing to talk politics whenever I needed to escape the computer. Karen Gormandy performed her usual magic with agent-y details. Kate McKay copy-edited out my typos, and Joe Gannon created the pleasing pages on which you are reading this. Each of these people had a hand in bringing *Leonardo's Foot* to life.

—*Carol Ann Rinzler*

# 1

# DESTINY

"Anatomy is destiny."

Sigmund Freud, "On the Universal Tendency
to Debasement in the Sphere of Love" (1912)

THERE ARE 206 BONES in the adult human body. When you put down this book, push back your chair, and stand up, you will be standing on fifty-two of them: your two feet.

Like virtually all creatures on earth, human beings are bilaterally symmetrical, with a relatively long, sometimes skinny central body to which limbs are attached in pairs, one of the pair on either side, even among creatures like centipedes who may have an odd number of pairs of legs, but always an even number of legs. The starfish and their relatives appear to be the only multicell creatures with an odd number of limbs, but even their limbs are evenly placed, this time around a circular body, an arrangement known as radial symmetry.

What makes us different is that unlike virtually all the others, we are bipeds. We naturally and consistently stand and move on two feet rather than four or eight or any other multiple of two. We share our bilateralism with practically everyone, but the only other bipeds are two distinct groups of mammals, the macropods—"big footed" kangaroos and wallabies—and the very small-footed kangaroo mice, jumping rodents native to the southwestern United States. Birds are also bipeds, but as avian anatomists know, birds hop on what looks like two feet but is actually comparable to our toes; the "heel" of a bird's foot is part of a toe; the thin piece just above that corresponds to the sole of the human foot. Theoretically, all of us can walk, run, jump, and jog, although the last–a compromise between walking and running—is pretty much the province of humans.

Every member of the human clan owes his or her two-foot stance first to the prehistoric sea creatures who developed fins strong enough to enable them to crawl up the banks of the local watering hole and become land animals; then to *Eduibamus cursoris,* whom the Carnegie Museum of Natural History (Pittsburgh) calls the world's oldest known biped reptile and whose 200,000,000 year old, slightly more than 10 inch fossil was discovered at a German quarry in 2000; followed by some dinosaurs; and finally to those birds who stood up on two legs and started to move by putting one foot in front of the other. And let's not forget the lucky confluence of geography, climate, and biological selection that contributed to the foot on which we stand today.

Leonardo da Vinci (1452–1519), no slouch himself at anatomical mechanics, described our foot as "a masterpiece of engineering and a work of art." His famous drawing, *The Vitruvian Man*, lays out perfect standards for the ideal human male figure from head to, yes, the toes attached to a foot, which ideally should measure one-sixth the height of the body.

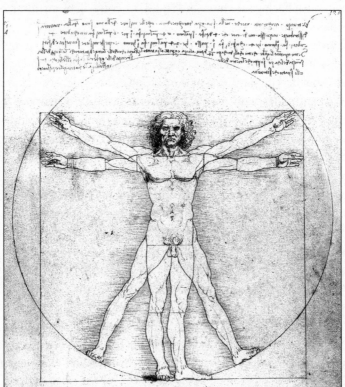

The measurements of the *Vitruvian Man* are based on the writings of Vitruvius (Marcus Vitruvius Pollio), the first-century Roman architect, engineer, and author of the classic 10-book treatise *De Architectura* (c. 20 BCE).

Vitruvius' passion was proportion, which he defined as "a correspondence among the measures of the members of an entire work, and of the whole to a certain part selected as standard." Vitruvius selected as the standards for the human male the height of a man or the height of the face. In a passage in Book 3, he wrote that "the human body is so designed by nature that the face, from the chin to the top of the forehead and the lowest roots of the hair, is a tenth part of the whole height; the open hand from the wrist to the tip of the middle finger is just the same; the head from the chin to the crown is an eighth, and with the neck and shoulder from the top of the breast to the lowest roots of the hair is a sixth; from the middle of the breast to the summit of the crown is a fourth. If we take the height of the face itself, the distance from the bottom of the chin to the underside of the nostrils is one third of it; the nose from the underside of the nostrils to a line between the eyebrows is the same; from there to the lowest roots of the hair is also a third."

Moving farther down the body, Vitruvius' other perfectly proportioned measurements include:

From the breast to the hairline: ⅙ the man's height.
The width of the breast: ¼ the man's height.
The root of the penis: at half the height of a man (this measurement is often omitted from the Vitruvian list).
From the armpit to the elbow: ⅛ the man's height.
From the elbow to the tip of the hand: ⅕ the man's height.
The length of the forearm (from elbow to wrist): ¼ the man's height.
From the wrist to the tip of the fingers: ¹⁄₁₀ the man's height.
From one fingertip to the other with the arms outstretched: equal to the man's height.
From the wrist to the tip of the fingers: ¹⁄₁₀ the man's height.
The foot: ⅙ as long as the man is tall.

With all this in mind, Vitruvius then laid out rules for the overall geometric diagram of the male body: ". . . if a man be placed flat on his back, with his hands and feet extended, and a pair of compasses centered at his navel, the fingers and toes of his two hands and feet will touch the circumference of a circle described therefrom. And just as the human body yields a circular outline, so too a square figure may be found from it. For if we measure the distance from the soles of the feet to the top of the head, and then apply that measure to the outstretched arms, the breadth will be found to be the same as the height, as in the case of plane surfaces which are perfectly square."

Take a ruler marked off in centimeters, measure a picture of the *Vitruvian Man*, and you will see that Leonardo drew the man exactly as Vitruvius proposed.

With maybe one interesting added detail.

Artists are not known for their modesty. In *Lives of the Most Excellent Painters, Sculptors, and Architects* (1550), Giorgio Vasari (1511–1674), the architect, artist, and author known for his important biographies of Renaissance painters and sculptors, described da Vinci as "an artist of outstanding physical beauty" and a man "endowed by heaven with beauty, grace and talent."

Dutch portraitist, illustrator, and activist Siegfried Woldhek (he once headed the Netherlands branch of the World Wildlife Fund and Dutch Birdlife International) believes that Leonardo used his own features as the face of the *Vitruvian Man*. In 2008, at a TED (Technology, Entertainment and Design) conference in California sponsored by the non-profit Sapling Foundation, Woldhek described how after combing through more than 120 male portraits by da Vinci he found three—*The Musician*, the *Vitruvian Man*, and the well-known drawing of an old man—in which the features conform to the descriptions of da Vinci's at ages 33, 38, and 63, the approximate ages of the three portraits.

In other words, the perfect face of the perfect *Vitruvian Man* may really be Leonardo da Vinci's.

## From there to here

Over the centuries since Vitruvius wrote his rules and da Vinci made them immortal, the marvelous human foot has walked its way into the English language.

For example, sometimes we mistakenly set off on *the wrong foot* when, to be successful, we should *put our best foot forward* and *start off on the right foot,* a trio of locutions that track back to the ancient relationship between left (*sinistre*) and right (*dextra*). By the way, if you think that many of the people in Egyptian paintings seem to have their feet on wrong, you're right. In Egyptian art, only the royals had both a right foot and a left foot;

lesser mortals were drawn with *two left feet*, then, as now, a descriptor for an inelegant person.

We say that timid people have *cold feet*, those with a hidden fault have *feet of clay*. The decisive will *put their foot down*, the bold *jump in with both feet* and, regardless of the fall, always *land on their feet*. Those in a hurry *hot foot it along*. If someone fails to give us a proper answer to a proper question, we will *hold his feet to the fire*. And leaving people who have treated us badly—or, in the more mellifluous words of Matthew 10:14, those "whosoever shall not receive you nor hear your words"—we *shake the dust off our feet*.

Finally, at The End, most of us leave this world carried out *feet first* so that our ghost will not be tempted to return. Unless the feet in question are attached to the body of a priest, one who is carried from the church head first, looking back at his congregation to which he expects to return . . . at least in spirit.

As we humans became the dominant species, it has been common to credit our progress to the evolution of our increasingly more complex brain with its highly specialized folds and grooves and multiple connections. In fact, we stood straight before we began to think straight, and in the process, our two feet with their fifty-two bones, sixty-six joints, and 200 assorted muscles and tendons have influenced not only our language, but also our culture, our politics, and of course our medicine.

To track how we got from there to here, begin by imagining yourself back in time, at a moment before history when you, an early hominin, are sitting on the ground, munching happily on berries—or as a team of scientists from the Max Planck Institute for Evolutionary Anthropology in Leipzig suggested in 2012, hard foods including grasses, grains, and tree bark. Your bottom is firmly flat on the ground; your trunk is perpendicular to the ground. Unlike four-legged mammals such as your dog or cat who sit with their front legs set straight from shoulder to ground to balance their bodies, your front legs—your arms—are free so that you can lift your food to your mouth rather than lowering your mouth to the food.

When you finish eating and are ready to move forward, you walk or run on two rear feet and perhaps the knuckles of your front feet, as apes and monkeys still do. Your big toe has not yet moved in close to the other four so above ground, in the trees, you can still use all four limbs with their opposable thumbs to grasp the branches, and your long arms to swing you from tree to tree through a canopy that looks nothing like the dreamy "murmuring pines and hemlock" Henry Wadsworth Longfellow created for the heroine of *Evangeline, A Tale of Acadie,* (1847), an epic poem that is one of the most popular works in American literature and the bane of school children who were made to memorize as many lines as they could cram into their heads. Instead, this forest was a tropical landscape similar to a modern rainforest that for tens of million years before you arrived on the scene, seemed to create a band all around the planet, skipping the oceans.

Now imagine that in the blink of an eye, evolutionarily-speaking, the canopy disappears; the forest, a place where more than 80 percent of the land was covered with trees, is gone. Instead you now inhabit a landscape known as the original savannah, a land of grass plus mini-forests, clumps of trees that take up less than half of any given area. Blink again, and the tree clumps are even more widely spaced in what is now a vast ocean of tall grass, the modern savannah that once covered nearly two-thirds of the African continent.

Anthropologists had long believed that the first savannah appeared less than 4 million years ago. But in 2011, geologists from the University of Utah traced chemical markers in the soil to describe and date the ratio of grass to trees in East Africa where fossils of the oldest known human ancestor have been identified, and concluded that savannahs existed 2 million years earlier than previously imagined.

Clearly, this change in the landscape was a problem for the "you" of this imagined scenario. Unable to digest grass, you suddenly find the food supply restricted to a less-plentiful store of fruit- and vegetable-bearing plants and trees. And to add insult to injury, at ground level, having once had your whole kingdom before your eyes, you can no longer see beyond your nose because the grass is taller than you are.

What to do? How to adapt?

Stand up.

Of course, it did not happen that quickly nor did we stand up just to look over the grass. Think of it instead as our moving step by step through the multiple frames required to make a Disney cartoon or as a modern pixel character moves his arm, and then multiply those frames over hundreds of thousands if not millions of years. Then at the end, multiply the frames yet again because as every toddler demonstrates, standing up is hard to do. To rise on two legs and then stay upright you must constantly make small adjustments to your balance. Walking or running is geometrically more complicated. You have to push off with your foot (actually the big toe), bounce back and forth on your ankles, and lift and extend and lower your knee, all in an unconsciously coordinated and perfectly smooth sequence that some have described as "controlled falling"—pause in mid-step and you are likely to trip and fall.

But once you master the trick, the rewards are obvious.

First, standing up makes you taller. On the ground, it is now easier to peer over what's in front of you and thus easier to check out what's happening around you. Standing up frees your arms to carry things; the hunter–gatherers can now lug home more prey or plants over longer distances than if they had to carry the booty on their backs or in their teeth, and the warriors can also fight more effectively with hands free to scratch, punch, or swing a club. In the searing African afternoon, you may be more comfortable because standing up presents more of your body surface to whatever cooling breezes may drift pleasantly by. Finally, standing up exposes the front of your body, making sexual display inevitable, which is likely to help to continue the species. But no advance comes free of some regret.

Once having stood up, your ancestors, the early hominins, pushed along toward full human-ness, perhaps by rewarding the person who made the best use of his (or her) ability to protect the clan, get the food, and use his (or her) freed arms to swat that faster sabre tooth tiger with a club. In other words, although our climbing the evolutionary ladder has always been attributed to our bigger brain, much anthropology and archeology tell us that the sturdy foot came first.

## The anthropology of an upright posture

Ask the next ten people you meet to describe an archeologist, writes Jeremy Sabloff, president of the Santa Fe Institute, and those who don't say *Indiana Jones*—or Harrison Ford—are likely to "describe a person digging in the middle of an excavation (probably dressed in khaki and wearing a pith helmet!) as dirt flies in all directions."

He's right. Archeologists really *are* people who like to play in the dirt and with rocks, preferably rocks with fossils inside. Even in an era of technological advances such as the proton magnetometer, a tool you move across the ground to sense any anomaly such as an iron spear or a fired clay pot or an indentation (think, *tomb*) underneath, eventually you have to dig up what's there, dust it off, chisel away the excess, and study what's left, hoping to find a new clue to how humans became or began to behave human.

Sometimes the rocks tell a tale no one expects, which is what happened to Raymond Dart.

Raymond Arthur Dart (1893–1988) was born in Brisbane, Australia on February 4, 1893 during a moment of high drama. The Brisbane River was flooding and, as *The Brisbane Courier* reported two days later, "the first Saturday of February 1893, has proved the most memorable in the calendar for many years past; we hope it will remain the most memorable for many years to come. On every vantage ground around Brisbane crowds gathered yesterday to witness the imposing and fearsome sight . . . Hundreds of wooden houses, once the happy homes of owner or occupier, careered upon the flood often remaining whole till they struck Victoria Bridge, when they crashed like matchboxes and broke away into shapeless masses of wood and iron . . . Steamers were driven ashore or laid on the tops of wharves." More to the personal point, the flood waters surged as high as second-floor windows, through one of which the attending midwife is said to have floated the newborn Dart and his mother out to a waiting rowboat.

Having survived the flood, Dart grew up on the family farm which he is reported to have loathed even as a child. He planned to escape as soon as possible, perhaps to serve as a medical missionary in China, but World War I intervened. Having earned two degrees in biology from the University of Queensland and two medical degrees from St. Andrew's College in Sydney, he enlisted in the Australian Army Medical Corps in 1917. Two years later he was discharged in London where fellow Australian Grafton Elliot Smith (1871–1937), an authority on anatomy and the evolution of the human brain and chair of anatomy at University College in London, soon hired Dart as a senior demonstrator, the British term for research or teaching assistant. Three years after that, Smith urged Dart to apply for the position of professor of anatomy at the new University of the Witwatersrand in Johannesburg, South Africa.

The teacher's suggestion did not please the student. Dart "hated" the idea of leaving London's libraries and laboratories to join a "new and ill-equipped University," complaining that he had "lived a pioneer's life for quite long enough in my younger [sic] days." But Smith prevailed, and in January 1923, just before his thirtieth birthday, Dart arrived in South Africa.

Once ensconced in his new position, he turned pragmatic. Now his aim was to build an anatomical collection of local finds such as the plethora of baboon skulls turning up at a limestone mine near a small city called Taung about twenty-eight miles west of Johannesburg. One day in September 1924, as he was getting dressed to serve as best man at a wedding to be held in his home, two new boxes of rocks arrived. With the groom literally tugging at one sleeve and his wife tugging at the other, Dart opened the boxes.

"A thrill of excitement shot through me," he wrote thirty-five years later in *Adventures with the Missing Link* (Harper, 1959). "On the very top of the rock heap was what was undoubtedly an endocranial cast or mold of the interior of the skull. Had it been only the fossilized brain cast of any species of ape it would have ranked as a great discovery, for such a thing had never before been reported. But I knew at a glance that what lay in my hands was no ordinary anthropoidal brain. Here in lime-consolidated sand was the replica of a brain three times as large as that of a baboon and considerably bigger than that of an adult chimpanzee. The startling image of the convolutions and furrows of the brain and the blood vessels of the skull were plainly visible. . . . But was there anywhere among this pile of rocks, a face to fit the brain?"

Later that day, having done his duty as best man, Dart returned and "ransacked feverishly through the boxes. My search was rewarded, for I found a large stone with a depression into which the cast fitted perfectly. . . . I stood in the shade holding the brain as greedily as any miser hugs his gold, my mind racing ahead. Here I was certain was one of the most significant finds ever made in the history of anthropology."

Over the next three months, Dart dusted and cleaned and scraped away, sometimes with homemade tools such as sharpened knitting needles borrowed from his wife. By Christmas day, he had removed enough debris to show the face of a creature he christened the Taung Child because its teeth were obviously those of a very young individual; eventually, the age of the fossilized creature at the time of its death was put at two or three years.

The Taung Child's brain was similar in size to that of an adult chimpanzee, but the skull was not an ape's—or at least not that of any known ape.

Gaul, Julius Caesar famously wrote, was divided in three parts. Your skull is divided in two. The top part is the cranium, a curved dome created by six bones: the ethmoid, sphenoid, frontal, parietal, temporal, and the occipital that fuse in the first 18 months of life to form a helmet that protects the brain and frames the eyes and ears. Under that is the mandible, the hinged lower jaw whose joints make it possible for you to open and close your mouth so that you can eat and drink and speak intelligibly.

Your cranium is round-ish, but not completely smooth because the muscles that move your jaws, head, neck, shoulders, and upper back are attached at specific points on the skull. On the ape or early hominin skull, these attachments produce significant bumps called ridges or crests. The sagittal crest on top anchors muscles that move the ape's jaws. The characteristic brow ridge that gives gorillas their fearsome appearance secures muscles that help relieve the stress of chewing and grinding hard plant food. And the nuchal crest on the occipital bone at the back of an ape or early hominin head is where the neck and back muscles that keep the ape (or hominin) head from falling forward are attached.

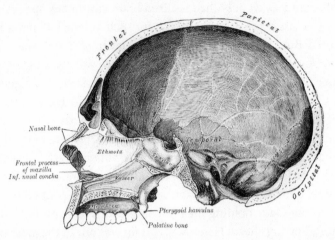

Some early hominins had various smaller versions of the sagittal crest, the brow ridge and the nuchal crest, but the fully human *Homo sapiens* skull has a flat forehead in place of a brow ridge, a minimal sagittal ridge, and an equally small site line where the nuchal ligament attaches. Because you stand upright, your head is supported by your spine. The nuchal ligament is a back-up system stretching from the bottom of the occipital protuberance (the bump you can feel at the back of your skull right above your neck) down to the vertebra prominens, the seventh and last bone in your neck. Its job is to help keep your head from rattling about too vigorously when you run or jump or otherwise bounce your body.

The foramen magnum is the opening at the back of the skull through which the spinal cord passes into the skull. In four-legged animals such as cats, dogs, horses, and cows, the foramen magnum is positioned so that the spinal cord forms a relatively straight horizontal line across the animal's back and into its skull. Like this:

But on the skull from the Taung quarry, the foramen magnum was further down and toward the underside of the head, similar to the positioning of the head and foramen magnum on a creature such as a Neanderthal/ Neandertal with a C-shape spine. Like this:

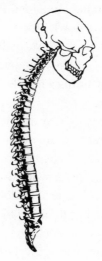

Dart's immediate conclusion was that the Taung Child stood upright, its back hunched and its head leaning forward, a step beyond the configuration of the four-legged animal's spine and head, but not quite the arrangement of our own Homo sapiens' S-shape thirty-three-vertebrae vertical spine with seven mobile cervical (neck) vertebrae; twelve thoracic (chest) vertebrae; five weight-bearing lumbar (lower back) vertebrae; five sacral vertebrae at the pelvis; and four caudal (tail) vertebrae that form the coccyx at the bottom of the spine. Like this:

Forty years later, the ape/early hominin C-curve and the human S-curve fit nicely into Rudolph Zallinger's well-known and often-imitated drawing, "Ascent of Man" in *Early Man* (Time Life, 1965), a simple, but memorable picture of how our ancestors slowly stood up to become the human race— or at least half the human race. In 2008, an illustrator for the magazine *New Scientist* tweaked the image slightly for an article about evolution "myths and misconceptions," substituting a female figure for the man standing latest in the evolutionary line.

Back in 1924, Dart was delighted with his Christmas day discovery ("I doubt if there was any parent prouder of his offspring"), which he named *Australopithecus africanus,* from the Latin *australis* meaning *southern* and the Greek *pithekos* meaning *ape.* A swift two months later, he published his findings in the February 1925 issue of the journal *Nature.* The Taung Child, he wrote, was more human-like than ape-like, and the position of the foramen magnum showed that he was a biped, walking upright with arms and hands free in front. In other words, the Taung Child, seemed an excellent candidate for the title of a missing link between apes and humans, one that would co-incidentally prove that humans appeared first in Africa.

It was a genuine *Eureka!* moment, and Dart waited expectantly for someone—or many someones—to shout an equally genuine, *Huzzah!*

But no one cheered.

Strike one was Dart's assumption that the Taung Child's small brain and semi-upright posture showed that the foot took precedence over the brain in moving us on our way to being human. The early twentieth century anthropology establishment, including the legendary Leakeys—Louis (1903–1972), Mary (1913–1996), Richard (1944– )—believed the opposite to be true, that the prime mover was our evolving brain, which fostered our increasing skill with tools.

The second strike was where Dart found his fossil. True, in *The Descent of Man* (1871) Charles Robert Darwin (1809–1882) had suggested that hominins appeared first in Africa, his evidence being the presence of chimpanzees and gorillas, the two species most like us. But the company line, endorsed by the leading European anthropologists as well as by the British anatomist Arthur Keith (1866–1955), a Pooh-Bah equal in stature to Grafton Elliot Smith, was that the ancestors of human beings, each with a large brain and a sloping ape-like face, had arisen in Europe. Others, such as Louis Leakey, voted for East Asia. Such small difference aside, all were sure that Dart had simply mistaken the skull of a juvenile ape for that of a hominin. Their argument was not unreasonable. Juveniles *do* often lack the facial characteristics of adult chimpanzees or gorillas.

Strike three? There was no bundle of similar skulls to back up Dart's claim; no more specimens had surfaced. Tuang was off the beaten anthropological path, and faced with a virtually universal lack of support, Dart and his students abandoned the excavation.

Like so many prophets, Dart was about to spend a lot of time without honor in his own land of anthropology.

Eventually, other biped hominins with chimpanzee-size brains did appear, such as the ape–man skull discovered by Mary Leakey in 1959 in Tanzania. But the woman who drove Dart's argument home was not Leakey, it was "Lucy," the small-brain, biped fossilized skeleton discovered near a lake in Ethiopia on November 24, 1974. Lucy was significant not only because she bolstered the theory of early bipedalism in Africa, but also because her fossil re-ignited an anthropological argument about whether hominins stood up first in the savannah or by a lake.

The "Aquatic Ape Theory" was introduced by marine biologist Alister Clavering Hardy (1896–1985) in an address to the British Sub Aqua-Club in 1960. Hardy's evidence was salty human sweat, which he believed showed that early hominins lived in a watery environment. Had they been walking on four legs, he explained, they would have drowned—or at least been seriously uncomfortable when entering the water in search of food. Hardy's idea of a waterside human predecessor turned out to be even less popular than Dart's presentation of the Taung Child as a missing link. Several more biped fossils, including Lucy's, were found near bodies of water, but the tale of the watery ape remains an anthropological outlier.

As for Raymond Dart and his foot-before-brain thesis, as Dennis O'Neil of California's Palomar College has written, "It is now clear that upright bodies and bipedal locomotion long preceded the large human brain. The early 20th century speculation that our ancestors would be large brained apes proved to be incorrect. We attained the full human form of bipedalism by about 2.5 million years ago, if not earlier. However, the size of our brain continued to increase in a punctuated evolutionary pattern."

Happily, although people still argue over where exactly *Australopithecus* fits or even whether he and she belong as an ancestor of Homo sapiens, Raymond Dart lived long enough to see his theories about bipedalism versus the big brain vindicated. Today the skull of the Taung Child sits in a box in the fossil room of the Institute for Human Evolution at the medical school of the University of Witwatersrand in Johannesburg, where both Dart and his discovery found a home.

## Determining destiny

In 1912, in his provocatively titled essay, "On the Universal Tendency to Debasement in the Sphere of Love," Sigmund Freud (1856–1939) wrote that "Anatomy is destiny" (*Die Anatomie ist das Schicksal*), thus igniting a fire whose flames he fanned with his later observation that the "great question that has never been answered, and which I have not yet been able to answer, despite my thirty years of research into the feminine soul, is 'What does a woman want?' "

His annoying befuddlement aside, Freud did have a point. The experience of being male *is* different from the experience of being female. So is the experience of being tall as opposed to being short or being right-handed rather than left. To side-step the gender issue and make the sentence more universal, all you need do is change Freud's comment from "Anatomy is destiny" to "Anatomy is *perception*."

Certainly, the anatomy that enabled us to stand up altered our perception of the world around us while changing both our about-to-be human body and our about-to-be bigger brain.

Consider, for example, the effect of our new and more adventurous daily diet.

As the forest receded and the savannah spread across the land, the evolving hominin's pantry expanded exponentially. Some believe that apes and very early man subsisted only on a high carbohydrate plant-based diet— not so. Our primate ancestors were familiar with animal protein. They had always consumed insects, which are plentiful in any forest. As Marvin Harris (1927–2001), an American anthropologist with a special interest in the history of food, often explained, people and animals, prehistoric as well as modern, tend to eat what's most easily at hand. Harris' metaphor for nutritional supply and demand was paper money. Suppose, he posits in *Good to Eat: Riddles of Food and Culture* (1986), that you live in a forest with twenty-dollar bills (think chickens) and one-dollar bills (think insects) clipped to the upper branches of the trees. Your first instinct is to reach for the twenties, but if there are only a few twenty-dollar bills and zillions of one-dollar bills that changes the equation. The one-dollar bills—in this case, the insects—win every time. And why not? Nutritionally speaking, bugs are prize worthy. In the raw, a 100 gram (3.5 ounce) serving of large grasshoppers provides twenty-one grams of protein and six grams of fat, the same amount of protein and twice the calorie/energy rich fat in a similar amount of raw chicken meat, minus the skin. As for the aesthetics, if lobster is your dish, please note that grasshoppers and lobsters both have a long skinny body with multiple legs. Both are good sources of protein and dietary fat. The only difference is culture, which often yields to appetite, intellectual as well as culinary, especially if you live in a tree and do not even have to leave home to pluck dinner off the leaves or bark.

Standing up makes a difference. With hands free, we can now hunt and kill and carry relatively easily. For Raymond Dart, this meant that the Taung Child and his relatives were hunters and probably violent killers, a belief then shared by many if not most of his contemporaries. "Man emerged from the anthropoid background for one reason only, because he was a killer," Robert Ardrey (1908–1980) wrote in *African Genesis* (1961). "Long ago, perhaps many millions of years ago, a line of killer apes branched off from the non-aggressive primate background. For reasons of environmental necessity, the line adopted the predatory way. For reasons of predatory necessity the line advanced. We learned to stand erect in the first place as a necessity of the hunting life. We learned to run in our pursuit of game across the yellowing African savannah. Our hands freed for the mauling and the hauling, we had no further use for a snout; and so it retreated. And lacking fighting teeth or claws, we took recourse by necessity to the weapon. A rock, a stick, a heavy bone—to our ancestral killer ape it meant the margin of survival."

Eventually, the explanation for the hominins' more meaty diet shifted from man as bloody killer to man as sneaky thief. As Robert J. Blumenschine, director of the Center for Human Evolutionary Studies at Rutgers University, has explained, wild cats and other hunting animals are likely to consume less than one-fifth of the bodies of their kill. That means early hominins could scavenge profitably and even subsidize a prehistoric food bank, leaving some scraps for other passing hungry creatures.

Either way, captured or stolen, the new higher protein diet changed our bodies, beginning with the face.

## The architecture of the biped body

Nature coddles no one, but she does protect her children, particularly the boys, by giving them weapons with which to defend their own progeny. Primate babies are born with undeveloped nervous systems, helpless as day old kittens. A chimpanzee infant clings to his mother for at least a year, a gorilla baby for three. Luckily, the adults in the family have a well-developed arms system: impressive canine teeth, up to four times larger in the male than in the female.

These teeth are not just for taking food apart; they are also for taking enemies apart, which makes you wonder whether all those toothy chimps in films and on TV are smiling or warning us to turn and run. In an interview with *Scientific American* in 1999, Knox College (Galesburg, Illinois) psychologist Frank McAndrew, an expert in facial expression, said it could go either way. "In the primate threat, the lips are curled back and the teeth are apart—you are ready to bite," he said. "But if the teeth are pressed together and the lips are relaxed, then clearly you are not prepared to do any damage." As we evolved from ape to human being, our facial expressions remained remarkably similar and seemingly hardwired. You don't have to learn how to smile or bare your teeth in threat, McAndrew says. "Kids who are born blind never see anybody smile, but they show the same kinds of smiles under the same situations as sighted people."

The normal modern adult ape or human mouth has thirty-two teeth: eight incisors (the big teeth in the middle of the upper and lower jaw), four canines (the pointed teeth on either side of the incisors), eight premolars (the teeth between the canines and the molars), eight molars (the

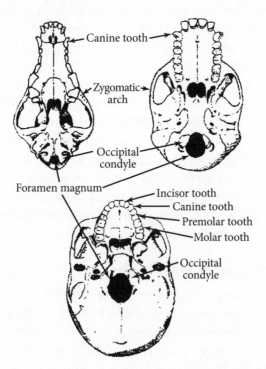

broad teeth in the back of the mouth). The ape jaw has room for another four molars; the smaller human jaw often does not. These "wisdom" teeth, so-called because they are the last permanent teeth to erupt, usually in the late teens as an adolescent becomes an adult and theoretically acquires wisdom, are usually extracted.

The *dental arch* is the term used to describe how teeth are set into the jaw. An ape's dental arch is a rectangle with one open side; the human dental arch is a rounded arc, narrower in front, wider in the back. Obviously, these arrangements affect the shape of the face, which is why fossil skulls are so valuable in showing how we went from ape to Homo sapiens.

The front of an ape's skull, his face, is prognathic (from the Greek words *pro* meaning *before* and *gnathos* meaning *jaw)*. His large jaws extend beyond the anterior cranial fossa, the forehead and front top of the skull that encloses the frontal lobes of the brain. Our face is orthognathic (from the Greek word *orthos* meaning *straight)*. Despite our projecting nose and chin, the human face is classified as "flat." The difference is due to the size of the canine teeth and how they are set into the jaw.

When Raymond Dart cleaned the Taung Child's face, he found canines smaller than an ape's, but larger than a human's. The jaws were similarly intermediate, projecting further than a human's, but less than a classic ape's. And the markings (patterns of wear) on the teeth suggested that the child had crushed rather than ground his food.

Apes grind very hard foods, such as nuts, whole, shell included, but moving the jaws from side to side like a mill grinding grain. Early hominins, like modern humans, crushed their meals of meat and other soft foods, exerting pressure simply by bringing their jaws together, bottom closing under top. Together, the smaller canines, the less protrusive jaw, the tooth markings, and the position of the Taung Child's foramen magnum pointing to a semi-upright posture, strengthened Dart's conviction that *Australopithecus africanus* was an intermediate species between apes and humans, more human in their dental arrangement but without the human chin.

All mammals have a lower jaw, the mandible, that begins as two separate bones, one left, one right, separated by a fiber and cartilage joint called a *symphysis*, Greek for *growing together*. In many animals, such as goats and sheep, the bones remain separate, allowing the two sides to flex independently, in a rolling grinding motion that often seems comical to humans (think of watching a goat eat). Some mammals, such as camels and hippos and horses, have mandibles whose bones fuse at about the time that their teeth appear; in apes and humans this occurs within the first year of life. But of all the roughly 5,000 species of mammals on earth, only one—us—has a mandible re-enforced by a bar of bone that forms a protruding bump, the mental (from the Latin *mentum* meaning *chin*) protuberance, at the center.

Explanations for the chin abound. Perhaps it stabilizes the lower jaw, reducing stress on the symphysis. Or it may increase the force with which we chew our food. Or it may enhance our ability to speak clearly. Or it may be sexually appealing; the male chin is more rounded and more prominent than the female, which is more delicate and pointy. Unfortunately, intriguing though these assumptions may be, there is a counter argument for each and every one. Goats, cows, and horses munch away on hard grains and apes chew through nut shells without a chin. As for speaking clearly, so long as you can open and close your mandible and move your tongue, you can make yourself understood. And that sex thing? Maybe. Maybe not. Either way, our chin remains an evolutionary mystery.

Not so the rest of our body, which, over the millennia, has clearly evolved to accommodate an upright posture.

Begin with the shoulders. Like an ape's shoulder joints, ours are still flexible; although as a very long list of baseball pitchers can tell you, the muscle-and-tendon rotator cuff around the shoulder is exquisitely sensitive to stress. Bipedalism has made some muscles once used to climb and swing redundant. For example, most of us no longer have a subclavius muscle stretching from the first (top) rib to the collarbone to stabilize the back and neck while walking on four legs. Some higher apes and humans no longer have one or more palmaris muscles running from the elbow to the wrist to strengthen the arm and flex the wrist and hand. Those who do have scored an anatomical bonus: Should they tear another, more important muscle, their surgeon can clip out the palmaris and use it to reconstruct the injured tissue.

Even without extra muscles, our arms can still support our weight in push- or pull-ups, but the humerus, the long bone in our upper arm, is not as robust as an ape's. The radius and ulna bones in our forearm are straighter, but not stronger than his, which are fused into one bone. Unlike an ape, whose arms are longer than his relatively short legs, our arms are shorter than our relatively long legs.

At the ends of the arms, our hands are smaller. Apes often swing themselves along. We walk, so while our fingers still grasp and hold efficiently, they are no longer curved like an ape's, nor as long as they were when our ancestors moved from tree to tree through the forest canopy.

Next up, or rather, down: our ribs. All apes have ribs in their necks; fewer than one in one hundred of us still do. About one in eight of us has an extra set of ribs, thirteen pairs like the chimps and gorillas rather than the standard twelve pairs for humans arranged in a rib cage whose shape was influenced by our stand-up diet. Chimps and gorillas do not have a waist because their rib cage, which looks like a bell—narrow at the top, wider at the bottom where it meets flaring hip bones—has no room for a waist separating ribs from hips. Our rib cage is shaped like a barrel with a clear separation, a waist between ribs and hips, a testament to our bipedal catch-all diet. Penn State paleoanthropologist Allan Walker, a member of the Richard Leakey team, which in 1984 discovered the skeleton of Turkana Boy, another African biped, wrote that the narrow-to-wide ape rib cage was designed to accommodate the long herbivore intestinal tube whereas the omnivore human intestines fit nicely into the shorter barrel shape cage.

Pulling nutrients out of a plant-based diet high in insoluble dietary fiber requires a long, multi-tasking intestinal tract. The higher the proportion of insoluble fiber, the more complicated the digestive apparatus will be. Herbivores such as cows that live on plants such as hay and grass use salivary enzymes to start the process of breaking the food apart into its nutritional components, but they still need more than one stomach to complete the job on that insoluble fiber. Animals that feed on softer fruits and vegetables have shorter intestinal tubes and only one stomach. Food from animals is even easier to digest, so carnivores such as the big cats get by with one stomach and an even shorter intestinal tract. We omnivorous human beings cannot digest insoluble dietary fiber, but our relatively short gut efficiently processes nutrients from both animal and soft plant foods.

Below the waist, our two-footed stance has changed the shape of the pelvis. Our ilium, the largest bone in the pelvis and the place where abdominal muscles attach, is nearly twice as large as a chimpanzee's, and it is wider and lower on the back, which has its own particular soft tissue. Monkeys and apes can sit up, but when they do their heads still lean forward because they lack the muscles needed to support the back and stabilize the hip and leg. Humans have three such muscles: the gluteus maximus, the gluteus medius, and the gluteus minimus, all packed tightly into our distinctively larger buttocks, a feature that separates us from most apes whose rear ends are generally flat. Maximus cushions us while sitting and helps us to turn or straighten our legs when hopping, jumping, or climbing. Medius, the middle, and minimus, the smallest, are both layered under maximus; they stabilize our upper body when we lift one foot, for example, while taking a step so that we do not tip over on the other one. "Old World monkeys," a group that includes baboons, macaques, mandrills, and any of the approximately one hundred species of monkey that live in Africa, along the Arabian coast of the Red Sea, and in Asia from Afghanistan to Indonesia, do have brightly colored sometimes padded buttocks and genitalia. But they don't have our strong gluteus muscles with their multiple connections to pelvis and legs, which as Robert Ardrey has observed, make possible "agility and all the turning and twisting and throwing and balance of the human body in an erect position," not to mention empowering a very large industry devoted to the care and coddling of the human lower back.

Because we stand up on our legs with our arms free, there's nothing surprising about bipedalism's influences on these limbs. Some, but not all of these changes, produce a better survivor, one who can catch dinner or outrun a more ferocious diner.

As noted earlier, the quadruped ape's arms are bigger and more powerful than ours. The long bones in the ape's forearm—the humerus, the radius and the ulna—may be fused so the arms are stronger, and overall often longer than the back legs. To facilitate staying up once we stand up, the long bones in our human legs (the femur above the knee and the tibia, and fibula below) are longer and stronger than the long bones in our arms (the humerus between the shoulder and the elbow and the ulna and radius between the elbow and the wrist). The ratio between the length of the forelimbs and the length of the back limbs is called the intermembral index (IM or IMI). The formula for the IM is written like this:

$$\frac{\text{length of the forelimb}}{\text{length of the hind limb}} \times 100$$

There are seven basic IM ratios. For convenience, they may be grouped into three broad categories which modern zoologists can use to describe prosimian and primate limbs and thus tell how the modern animal moves through its environment; when examining fossils, anthropologists can use these measurements to speculate on how these long-ago creatures moved.

An IM equal to or close to 1 describes an animal whose front and back legs are of equal length. Primates and prosimians whose short arms and legs are equal in length walk on four legs on the ground and across tree limbs. Those with long limbs equal in length walk on four limbs, often on their toes. Animals with *very* long limbs that are equal in length climb and swing with all four limbs.

An animal with an IM greater than 1 has arms longer than its legs. These animals are likely to move by brachiating (from the Latin word *bracchium* meaning *arm*), swinging arm over arm through the trees. Chimpanzees and gorillas have arms longer than their legs and may also brachiate, but more commonly they are quadrupeds, walking on their rear feet and the knuckles of their front legs (orangutans, which are also quadrupeds, fold the front foot into a fist and walk on that rather than on the knuckles).

Finally, primates whose rear legs are longer than their arms have an IM lower than 1. These animals move by leaping from place. The primary primate exception to this rule, of course, is us, an entire species that walks erect on two firm and steady longer legs. In their summary of comparative limb structure published in the *Journal of Anatomy* in the summer of 2000, Bernard Wood and Brian Richmond of George Washington University called this inequality of limb length the most striking difference between apes and humans: "The great absolute and relative length of modern human lower limbs . . . increases stride length and thus the speed of bipedal walking."

An important requirement for standing up safely on two legs is a skeleton that keeps the body balanced over the legs. Like apes, early hominids moved on four limbs, so they always had at least two limbs on the ground at once to balance their bodies. Ours is a more delicate task because our hips are wider apart. If our legs were also wide apart, we would have a hard time to keep from falling over. Our femur has evolved to an angle that brings the knee joint in closer to the middle of the body, an orientation known as *valgus knee* ("knock knees" are an extreme version of this orientation). With the knee in closer, our feet are positioned under the center of our body so we can stand up, as Lucy is believed to have done. Wood and Richmond write that her "bones are rife with evidence clearly pointing to bipedality." Her femur is angled correctly; her knees are set with a prominent edge on the top to keep the knee from slipping out of place; the condyles (ends) of the bones are large enough to support the extra weight of standing on two legs instead of four; and her pelvis is wider and more stable. "The entire structure," they conclude, "has been remodeled to accommodate an upright stance and the need to balance the trunk on only one limb with each stride." As for the spring in that stride, we owe our bounce to our Achilles tendon, the tissue that ties our calf muscle to the bones in our heel and is vital for long distance running. Apes have no Achilles tendon; their calf muscles stretch right down to the bones in their feet.

Which brings us back, at last, to the fifty-two bones with which we began: our feet, two long, firmly arched pads with five perfectly adapted toes.

Start at the ankle, set squarely on our heel. Unlike a chimpanzee's ankle, which flexes sufficiently to permit the foot to twist around a tree while climbing, ours is designed to keep the long bones in the lower leg straight over the foot, an appendage whose evolution has resulted in three unique characteristics: the position of the big toe, the firmness of the arch(es), and the full fall of the sole on the ground.

Unlike other primates, whose feet are as flexible as their hands with an opposable thumb-like first toe, our foot is relatively rigid with three distinct arches firm enough to absorb the shock as our foot strikes the ground and the force of our body's weight crashes onto our feet. We are not the first to have had these arches, or something like them. In September 2011, Carol Ward of the University of Missouri School of Medicine and William Kimbel of Arizona State University reported that a metatarsal bone found with Lucy's fossilized skeleton showed that Lucy, whom Ward calls "the poster girl for her group of ancient hominins," had arch-stabilized feet and was perfectly capable of standing up and walking, though probably not as straight up or as efficiently as we do.

Farther forward on the foot, our toes are no longer fingers. They are shorter and closer together, and our big toe (the *hallux,* from the Latin *allex* meaning *big toe*) is not just another digit. Unlike the long first toe on an ape's rear foot, ours is short, straight, and adducted, that is, moved in close to our other four (noticeably shortened) toes to help us balance when we stand and to lift and push our foot up and forward with each step. Together, the five toes absorb pressure as we walk or run.

Underneath our ankle, toes, and arch is the bottom of our foot, the sole, the final platform on which our body rests and by virtue of its shape and scent, for some an object of intense desire, about which more later. For the moment, it is enough to observe that we human beings are not only bipeds, we are *plantigrade* (from the Latin words *planta* meaning *sole of the foot* and *gradus* meaning *step* or *walk).* We stand on the whole foot, not on our heel or on the ball of our foot or on our toes. We are, quite simply, balanced on our feet.

Even now, it is widely believed that *Homo sapiens* owes its primacy as a species to an increasingly more complex brain. But here, as is so often the case, the common wisdom is wrong. What changed us first was not the centuries-long evolution of our brain, but the transcendent moment when our hominin ancestors the Taung Child, Lucy, and all their friends and neighbors stood up on the African savannah or maybe next to an African stream. At the time, the brain that ruled their bodies was larger than an ape's, but smaller than a human's. Standing up immediately gave them entry to a new and larger universe brimming with a richer, more varied, higher protein diet, whether fish by the water as Alister Hardy proposed or flesh on land as most others accepted.

Either way, the new menu provided the energy required to enlarge the brain. With that bigger brain, as Robert Ardrey wrote, "We became man." And walked into the future on our own two feet.

# 2

# DISABILITY

"A true critic ought to dwell rather upon excellencies
than imperfections."

Joseph Addison, *The Spectator* (1711)

UNLIKE LEONARDO'S perfect Vitruvian foot, whose length from heel
to toe equaled exactly one-sixth the height of the person to whom it was
attached, our own feet are usually longer or shorter or wider or more narrow than the Roman ideal.

They may even be twisted out of shape. Every year around the world,
about one in every 1,000 babies is born with a clubfoot, that is, a foot
twisted up or down so that the sole cannot be placed flat on the floor.
Clubfoot is our most common congenital lower limb deformity. In history—both real and mythological—it has tormented gods, kings, and com-
moners alike, so frequently in fact that even in the ancient world whose
culture and law sanctioned a parent's disposing of an imperfect infant,
this one deformity was not a death sentence.

The Greek poet-philosopher Hermodorus of Ephesus (c. 6$^{th}$–5$^{th}$ centuries BCE) is a bit player in this story, simply a man who carried a message
from one town to another, unremarkable except for the fact that the message came from Plato and the town to which Hermodorus carried it was
Rome where he helped to convert the ancient policy of dealing harshly
with deformed infants from Roman custom into Roman law.

One day sometime around 459 BCE, Hermodorus did what philosophers
usually do: He told other people how to behave, in this case proposing citi-
zenship for freed men in Ephesus and the right to public office for their
children. The very thought so offended the City elders that they exiled him.

Hermodorus' friends were not amused.

His fellow philosopher Heraclitus (c. 535–475 BCE) said that the Ephesians "all deserve to have their necks broken as they grow up, so that the town should be left to minors because they drove away his friend Hermodorus, the best of them all, and gave as their reason for so doing that amongst them none should be more excellent than the rest; and if anyone were so, it should be elsewhere and amongst others."

After which Heraclitus himself was tossed out of town "for misanthropy and refusal to smile."

Several centuries later, Ephesus was the capital of Proconsular Asia, the Roman province comprising western Asia Minor, the peninsula between the Black Sea and the Mediterranean, a major transport center for goods from both Greece and Rome with roads leading to all the major cities along the way. Around the year 51 CE, Saul (5–c. 67 CE), who later became the disciple Paul, stopped off for a short visit on his way from Greece to Syria. Three years later, he returned to Ephesus. He was not pleased with what he found.

There was, for example, the matter of the Temple of Artemis (Diana), the Ephesians' greatest monument. The temple was the largest open air theater in the contemporary world with a capacity of 50,000 spectators, just 7,545 fewer than the official capacity (including seats and standing room) of Yankee Stadium in New York. The newly named Paul opposed gods made by man, such as those embodied by the statues of the Greek and Roman deities. But the Ephesians said their statue of the goddess Artemis had fallen from the sky and was therefore divine. Besides, the temple provided asylum for debtors, a practice the Ephesians hotly defended before the Roman senate.

Clearly annoyed, Paul penned two scathing missives in which he berated the Ephesians for their obeisance to Cybele (goddess of nature and fertility, both linked, however delicately, to sexuality) and Dionysus a.k.a. Bacchus (the god of wine and celebration), as well as for their lust, lies, fraud, arguments, robbery, unnecessary litigation, their sexual behavior, and their decision to abandon their principles in favor of whatever the current ruling class ruled. He was equally contemptuous of the Ephesians' wealth because he believed that fate punishes not by taking away a person's possessions, but by granting the abundance that gives people the means by which to demonstrate their inevitable corruption.

Meanwhile, Hermodorus had long ago made it to Rome where he served as an advisor to the Decemvirate (from the Latin *decemviri* meaning *ten men*), the council charged in 455 BCE with creating the first written Roman code of law designed to deal even handedly with the rights of both the privileged and the less so. The Decemvirs began with ten bronze tablets of rules and regulations, but there was obviously need for more because a second commission, named five years later, filled another two with social and legal niceties such as the prohibition of marriage between members of different social classes (Table XI) and the promise that "whatever the People has last ordained shall be held as binding by law" (Table XII). With those additions, *The Twelve Tables* served Roman justice until 387 BCE when the invading Gauls sacked the city and destroyed the physical tablets. The Tables codified Roman law as follows:

Table I: Courts and trials
Table II: Trials
Table III: Debt
Table IV: Rights and duties of fathers
Table V: Guardianship and inheritance
Table VI: Ownership
Table VII: Land rights
Table VIII: Legal rules on injuries
Table IX: Public law
Table X: Sacred law
Table XI: Matters not previously covered
Table XII: Matters not previously covered

Perhaps the most troublesome of the Decemvirs' new laws were the ones in Table IV, which, with Hermodorus' advice, declared it to be a father's duty to make sure that "if a child is born with a deformity he shall be killed."

At the time, abandoning or killing deformed infants was accepted practice all across the Mediterranean world. Spartans, for example, already had such a custom, based on the laws promulgated by Lycurgus, who ruled Sparta sometime between 800 and 630 BCE—no one is absolutely certain of the dates. Some historians even consider him a legend rather than a real man, but still credit him with creating the city-state's constitution.

In any event, disabled Greek and Roman newborns who survived into childhood or even managed somehow to make it into adulthood faced a life considered less than valuable. So much less in fact that the "games" at the Coliseum included tossing these children under the hooves of galloping horses and setting matches between blind adult gladiators or between blind and sighted ones.

This callous disregard for the disabled was not limited to the secular society. In Leviticus 21:17–23, the Higher Authority was pretty clear when he told Moses what to tell Aaron, the first priest of Israel, about who could do what in sacred places: "Whosoever be of thy seed throughout their generations that hath a blemish, let him not approach to offer the bread of his God. For whatsoever man he be that hath a blemish, he shall not approach: a blind man, or a lame, or he that hath anything maimed, or anything too long or a man that is broken-footed, or broken-handed, or crook-backed, or a dwarf, or that hath his eye overspread, or is scabbed or scurvy, or hath his stones crushed; no man of the seed of Aaron the priest that hath a blemish, shall come nigh in order to offer the offering of the Lord made by fire; he hath a blemish; he shall not come nigh to offer the bread of his God. He may eat the bread of his God, both of the most holy, and of the holy. Only he shall not go into unto the veil, nor come nigh unto the altar, because he hath a blemish; that he profane not my Holy places; for I am the Lord who sanctify them."

Why were these people so careless with their imperfect, but helpless infants? There are several possibilities from the divine to the prosaic.

In pre-scientific society, where decisions were based on signs and symbols, a visibly deformed child signaled the displeasure of the gods and was therefore a threat to the family, the tribe, and the city. In a world where wars were fought primarily in hand-to-hand combat, the need for a strong and able cohort was obvious. In an agrarian society, the imperfect baby was one more mouth to feed, but one less hand (or foot) to work.

Then there was the thorny question of exactly when a human became human. Hippocrates believed that the human soul was present from the very beginning, in modern terms, from the moment of fertilization. Others thought otherwise, questioning whether a newborn was really a human being at all. Contrasting the human with the not-quite, Aristotle (c. 384–322 BCE) wrote in *The History of Animals* that a fetus begins with a "nutritive" soul similar to that of plants, moves on to a "sensitive" soul like an animal's, and then finally develops a "rational" or intellectual soul. This theory, known as *delayed hominization*, was adopted as Catholic dogma in 1312 at the Council of Vienne in northern France. Thereafter, Catholics, while opposing abortion, did not consider it to be murder until 1588 when Pope Sixtus V issued the *Bull Effraenatam* [*without restraint*] directing that anyone who performed an abortion at any stage of pregnancy would be excommunicated and subject to whatever punishment the civil law prescribed for murderers. Three years later, Gregory XIV overturned this, limiting the punishment to excommunication and that only for the abortion of a "quickened" fetus.

For some, however, neither quickening nor the acquisition of an intelligent soul was enough to guarantee human-ness. Like Hippocrates, Aristotle believed that a male fetus *looked* human 40 days into a pregnancy and a female at 80–90 days (one way to justify the idea that men were naturally stronger and more intelligent than women). But he also thought that despite appearances a newborn was not *fully* human until one week after birth, thus conveniently providing a seven day window in which to dispose of severely disabled infants. Among the ancient Jews, the waiting period was thirty days, although a male child was named seven to eight days after birth when he is circumcised, marking the moment when a Jewish identity passes from father to son. In some parts of Japan, a child became a human being when its first cry told everyone that the infant's spirit had arrived. Elsewhere naming became the important ritual, and eventually, for Christians, baptism.

However, even being labeled "human" was not enough to protect all infants. Sometimes even healthy children were seen as threatening. Egyptians considered twins evil children of the gods; twins of opposite gender were thought to have committed incest in the womb. Well into the Middle Ages, a visible disabling birth defect such as a missing limb was reason enough to eliminate the theoretically useless child, and disabled adults or those afflicted with disfiguring diseases, most commonly leprosy and small pox, were regarded as sinners and scapegoats for natural disasters and catastrophes ranging from the periodic visitations of the plague to earthquakes and drought.

The legal and social disqualification of the disabled was reinforced by art. As the often flat, one-dimensional Medieval portraits gave way to the richer paintings of the Renaissance, artists such as Titian, Donatello, Tintoretto, Botticelli, Caravaggio, and Raphael sought to revive the Greek and Roman ideals of perfectly proportioned human beauty so clearly depicted in Leonardo's *Vitruvian Man*, resulting in yet another reason to reject those who did not measure up—with the exception of those with a clubfoot, which from the beginning seemed to be a deformity with a difference.

## First impressions, catalogues, and curiosities

If your leg and foot were one long L-shape bone, you would never need a knee replacement or an ankle brace. Of course, you would have to lift the entire leg to move, so your ability to turn quickly would be severely compromised, and if you fell and broke the bone, you would have to replace the entire leg, thigh to toe. Luckily, as American architect Louis Sullivan once observed, form follows function (what Sullivan really said was "form ever follows function," but removing one word made the point faster and neater, so it stuck).

Our anatomy agrees. We have joints at the knee, at the ankle, and in the foot itself that make it easy to move forward, backward, and sideways. The downside, of course, is that we risk knee and ankle injuries and arthritis, as well as clubfoot.

The Greeks called clubfoot *strephenopodia* from the words *streph* meaning *twist* and the obvious *pod*. The name was perfectly descriptive. Congenital clubfoot is a dislocation either of the tibiotalar joint that connects the tibia (the larger and stronger of the two bones in the lower leg) to the ankle (talus) or the talocalcaneonavicular joint that connects the ankle bone to the heel (calcaneus) and to the navicular bone, one of the small tarsal bones on the top of the foot.

The basic medical name for congenital clubfoot is *talipes* (*talus* plus *pes*). A foot twisted down from the ankle is called *talipes equinus*, from the Latin *equus* meaning *horse*. In *talipes varus* (Latin for *bent*), the foot is turned toward the other foot. In the less common *talipes valgus* (Latin for *twisted outward*), the foot is turned away from the other foot. If the twisting is severe, an untreated child may walk either on his toes or on the outer edge of the foot. Either way, the deformed foot does look something like a horse's hoof. *Talipes calcaneus* is a second form of clubfoot. Here the foot is bent up, with the toes pointing toward the knee and the heel towards the floor. Normally, a person walks by putting the toe or the heel, then the rest of the foot, down on the floor. In talipes calcaneus, only the heel hits the floor. In either case, if the twisted foot is not treated fairly quickly, its muscles and tendons begin to contract to hold the twisted foot in its unnatural position.

Given the fact that birth defects were often regarded as a sign of worthlessness or worse, a person with a congenital clubfoot might well have tried to blame the problem on an injury. For example, Charles-Maurice de Talleyrand-Périgord (1754–1838) claimed his clubfoot was caused by his having fallen off a chest of drawers as an infant. The accident, he insisted, had crushed his foot which then hardened into a round mass. Medically, this is highly unlikely. Besides, biographers note that Talleyrand's mother disliked him as soon as he was born, suggesting that she was reacting to the sight of a congenital abnormality unwelcome in her eighteenth century aristocratic family. It might also explain why the family packed the boy off as quickly as possible, first to boarding school and then—because he was unfit for military service—to a seminary to train to be a priest, and eventually to become a bishop in keeping with his family's status. Talleyrand resented this mightily; the exile is said to have turned him hard and merciless, just the attributes he required to serve as Grand Chamberlain to Napoleon, to avoid the guillotine by siding with the revolutionaries during the Revolution, and generally to come out on top in virtually every political situation.

Later in life, clubbing may follow an infection or an illness. Joseph Goebbels (1897–1945), the Nazi Reich Minister of Propaganda, is believed to have acquired his clubfoot as a complication of osteomyelitis, an infection inside a bone. Afterward, Goebbels walked with a leg brace and that, along with his puny 100 pound physique led to his being rejected when he sought to enlist in the German armed services during World War I. Sir Walter Scott (1771–1832) acquired his clubfoot after a bout with polio. Scott's recollection of his illness appears in John Gibson Lockhardt's seven volume *Memoirs of Sir Walter Scott* (1808):

> "I showed every sign of health and strength until I was about eighteen months old," Scott recalled. "One night, I have been often told, I showed great reluctance to be caught and put to bed, and after being chased about the room, was apprehended and consigned to my dormitory with some difficulty. It was the last time I was to show much personal agility. In the morning I was discovered to be affected with the fever which often accompanies the cutting of large teeth [teething]. It held me three days. On the fourth, when they went to bathe me as usual, they discovered that I had lost the power in my right leg. . . . when the efforts of regular physicians had been exhausted, without the slightest success. . . . the impatience of a child soon inclined me to struggle with my infirmity, and I began by degrees to stand, walk, and to run. Although the limb affected was much shrunk and contracted, my general health, which was of more importance, was much strengthened by being frequently in the open air, and, in a word, I who in a city had probably been condemned to helpless and hopeless decrepitude, was now a healthy, high-spirited, and, my lameness apart, a sturdy child."

The first depictions of clubfeet appear to be in Egyptian tomb paintings, which is not surprising considering the consequences of the marital histories of the Egyptian royals. Pharaohs were believed to be children of the gods. Likewise, Roman emperors were considered celestial individuals temporarily on earth, and European royalty claimed appointment by divine intention. In each case, to keep the bloodlines godly, and the fortune in the family, the royals frequently married each other.

Unfortunately, what pleased the gods and the accountants did not necessarily result in healthy babies. In Europe's ruling houses, multiple intermarriages produced numerous genetic problems. Among the Hapsburgs of the House of Austria who claimed the Spanish throne in 1516, nine of eleven marriages during a 200-year reign were between blood relatives. One highly visible result was the Hapsburg jaw, a prominent mandible (lower jaw) sometimes pushed so far forward as to make it difficult if not impossible to bring the teeth together to chew solid food. There were also kidney abnormalities, mental retardation, impotence, and a sky-high rate of infant mortality. It is estimated that just half of the children born to the Spanish Hapsburgs made it alive to his or her first birthday; the end of the line came in less than two centuries. In the British House of Hanover, Queen Victoria's recessive hemophilia gene, carried by at least six of her female descendants, is known to have struck nine of her male children, grandchildren, and great-grandchildren including history's best-known hemophiliac, the ill-fated Alexei Nikolaevich, Tsarevich of Russia, son of Victoria's granddaughter Alexandra.

Geneticists generally consider marriages between relatives closer than second cousins to be potentially problematic. The Europeans pretty much stuck to relationships no closer than first cousins, but even that was a problem because children of such unions inherit more than six percent of their genes as identical copies, one from each parent. The Egyptians, on the other hand, sanctioned matches between siblings, half siblings, or in a pinch, father and daughter or grandfather and granddaughter. (Mindful of this history, modern Egyptian law prohibits marriages among persons whose relationships is closer than that of first cousins.)

Tutankhamen (c. 1325–1342 BCE), whose parents were brother and sister, was born into the much intermarried 18$^{th}$ Dynasty, one of the three clans that ruled from 1570 to 1085 BCE during the period known as the New Kings. For nearly 100 years after his tomb was discovered and opened in 1922, it was assumed that King Tut had expired after being hit on the head with a blunt instrument, a theory supported by the fact that there was a hole in the mummy's head. But in the winter of 2010, when a multinational team of geneticists began testing the DNA of the mummies in Tut's tomb, including two infants believed to be his daughters with his half sister Ankhesenatmun, the daughter of his stepmother Nefertiti, their tests showed not only that most of the people buried with Tut were related, but also that Tut had been a victim of haphazard consanguinity, suffering not one but at least two genetic strikes. Like his father Amenhotep IV, he had a cleft palate, and like his grandfather Amenhotep III, he had a clubbed foot.

Clearly, the Egyptians, like the Greeks and Romans, had no problems dispensing straightaway with infants born with birth defects. But their royals, after all, were god-like, so who knows? Maybe the populace just decided that if a deformity such as clubfoot showed up in more than one member of the semi-divine ruling royal family, the problem was simply part of the royal package. The idea is pure speculation, but if so, would have bought Tut's grandfather and his father their own rule and given Tut himself time on the throne until he died, perhaps of malaria. Or because Tut's mummy had a broken leg, along with the cleft palate and clubfoot, maybe one day, lame and walking with canes, he tripped and fell and then, absent sterile surgery and antibiotics, developed a galloping infection that carried him off at the tender age of 19, a romantically early death that whatever the cause ensured that he would forever be remembered as the "boy king."

He was by no means the only one in the neighborhood. Tiberius Claudius Caesar Augustus Germanicus, a.k.a. Claudius I (c. 10 BCE–54 CE), the fourth emperor of Rome and the first to be born outside Italy (he came from Gaul in what is today Lyon, France), not only had a clubfoot, he stammered. Widely considered a weakling, once in power as Emperor, Claudius proved the common wisdom wrong. He ruled for 13 years from 41 to 54 CE, during which time he conquered and colonized Britain and streamlined the administration of Rome itself. Then, like so many of his generation, he left office under suspicious circumstances, probably having been poisoned by his wife Agrippina, whose son Nero—Claudius' stepson—took his place.

Catalogues of human monsters, including monstrous infants, date as far back as the Babylonians. These early records were random collections of abnormalities, ranging from the fantastic (races of men with animal heads) to the realistic (Egyptian and Sri Lankan representations of achondroplastic dwarfs). *The Etymologies*, compiled by Isidore of Seville (c. 560–636) sometime during the years 615 to 630, was different. Its author, later canonized as St. Isidore, was born into an orthodox Catholic family in Cartagena, orphaned as a child, and raised by his brother Leander, whom he succeeded as Bishop of Seville in 600. Isidore was a prolific writer; *The Etymologies*, his magnum opus, was a 51-volume encyclopedia designed to explain all branches of human knowledge, including human diversity (Volume XI), based on the meaning of words and names.

Semantics aside, Isidore's important innovation was to group "monstrosities," including children born with birth defects, into twelve distinct categories:

Excessive growth (i.e., giants)
Insufficient growth (i.e., pygmies and dwarves)
Unusually enlarged body parts
Extra body parts (i.e., a sixth finger or toe)
Missing body parts (i.e., a missing bone in leg or arm)
Animal parts (i.e., webbed fingers or toes or excessive hairiness)
Delivery of a newborn animal rather than a human child
Misplaced organs or body parts (i.e., eyes too wide apart)
Unusual aging (progeria)
Composite creatures (i.e., the centaur or the mermaid)
Hermaphrodites [infants with apparently male (Hermes]) and
    female (Aphrodite) genitalia]
Monstrous races [i.e., entire nations of men such as "the Cyno-
    cephali ... so called because they have dogs' heads (*cyno* is
    Latin for dog, *cephalo,* for head) and their very barking betrays
    them as beasts rather than men. These are born in India."]

*The Etymologies* remained the standard reference for centuries. Although as time passed, the approach slowly–very slowly–became less fantastic and more scientific, Isadore's carefully constructed categories created a template for later categorizations of human abnormalities such as those defined by Fortunio Liceti (1577–1657) and Ambroise Pare (c.1510–1590).

Liceti appears to have been a natural overachiever. He graduated from the University of Bologna with doctorates in philosophy and medicine and is the author of two major medical treatises. In *De monstruorum causis, natura et differentiis* (1616) he addressed genetic anomalies; in *De spontaneo viventium ortu* (1618) he affirmed the spontaneous generation of living beings. Although both essays reflect the unscientific tradition of his day, he did have one foot halfway into the medical future, arguing consistently against the idea that birth defects were caused by divine wrath and searching, albeit in vain, for a physiological explanation.

Pare, the royal surgeon to the French kings Henry II, Francis II, Charles IX, and Henry III who ruled, one after the other, from 1547 to 1574, did the usual cataloguing. Intellectually, he was several steps ahead of his contemporaries in believing that birth defects were due to a "lack of discipline in the parents' seed," a crude, but intuitive suggestion that they were due to human malfunction rather than divine intent. On the other hand, his theories about the cause of clubfoot were classically naive. He thought the problem arose from the mother's sitting cross-legged while pregnant or having held the baby too tight against her body while nursing. His advice on treatment was also classic: He endorsed the method of the master, Hippocrates.

The Egyptians may have been the first to draw a picture of a clubfoot, but Hippocrates was the first to produce a written description of the condition, sometime around 400 BCE. Given the medical surround in which he worked—no anesthetic, no antiseptics, no diagnostic tools save fortune telling—the treatment he proposed was pretty much on target.

He advised beginning with gentle manipulation of the foot, then securing it with bandages to stretch the soft tissues and allow the foot to fall into a natural position in which the sole would be flat on the floor. Once the foot was overcorrected, new coverings were used to reinforce the correction and keep the foot from slipping back into the abnormal position. It is an extraordinary compliment to his skills as a physician operating strictly on observation and intuition that 2,000 years later Ambroise Pare endorsed as his own most of Hippocrates' treatment for clubfoot. Pare did introduce a boot of turpentine-softened leather with a wedge sole to hold the foot in place long enough to increase the chance of permanent straightening, but he understood that correcting the foot took time and patience. In "The defect called varus and valgus," one chapter in *Oeuvres* (1585), the multi-volume summary of his own career and accomplishments, Pare explained that when treating a clubfoot "one must not make *varus* and *valgus* children walk until the joints are well strengthened, so that they do not dislocate. And when one wants to make the children walk, one should split open some high shoes, little half boots, and laced up at the front, or fastened by little hooks: they should be of hard leather so as always to keep the bones firmly on the joints, and so that they have to stay there. And the sole must be higher on the side where the malformation will be inclined to turn, in order to force it to turn back to the necessary side."

This adaptation of Hippocrates' simple but effective approach worked well enough to keep it in vogue for centuries more—with occasional embellishments, not necessarily for the better.

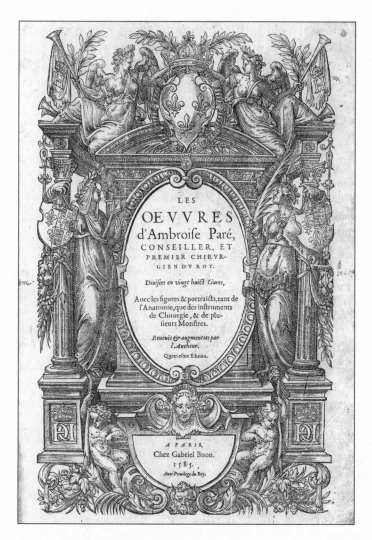

## Cruel and unusual cures

Italian anatomist and surgeon Antonio Scarpa (1752–1832) was a man whom one biography describes as authoritarian and "picky" with few friends and many enemies—rather like Tosca's nemesis Scarpio without the music.

He was also a man of many interests. His first success came in 1772 with a treatise on the structure of the ear. He is credited with creating the specialty of orthopedic surgery. His 1803 essay *Memoria Chirurgica sui Piedi Torti Congeniti dei Fanciulli, e sulla Maniera di Corregere questa Deformita* (*A Memoir on the Congenital Clubfeet of Children, and of the Mode of Correcting that Deformity*) was the first accurate description of a

clubfoot. However, unlike Hippocrates and Pare, Scarpa insisted on forc-
ibly straightening the foot, then fitting the patient with a multistrap rigid
boot known as Scarpa's shoe to hold the foot in position. No one else was
able to show that his method worked, so it was soon abandoned.[1]

Before the punishing regimens and devices prescribed by Scarpa
and some of his contemporaries were discredited, they claimed a host
of young victims including George Gordon, later George Gordon Noel,
6th Baron Byron, better known as the romantic poet Lord Byron (1788–
1824). Byron was born "with lameness and a twisted foot." As a child, he
was treated first by a man named Levander, generally identified only as
a "trussmaker for the local hospital," whose therapy consisted in rubbing
the boy's foot with oils, roughly twisting it into normal position, and then
locking the foot and leg into a "wooden machine" to hold it straight.
When this did not work any better than Scarpa's shoe, Levander was sent
away, and Byron's mother took her son to London to see the Scottish
physician Matthew Baillie (1761–1823).

Baillie's credentials were impeccable. He wrote the first British work
on pathology as a distinct discipline in 1793 (*Morbid Anatomy of Some of
the Most Important Parts of the Human Body*). He was the son-in-law of
Thomas Denman (1733–1815), author of a classic textbook on obstetrics.
His uncle was the anatomist John Hunter (1728–1793), regarded as the
father of experimental pathology. Baillie's treatment was orthodox, and
better yet, relatively effective. Using gentle manipulation and a special
boot, he was able to straighten young Byron's foot so that he could walk
easily, although with a slight limp.

As a result, Byron did manage to overcome his disability and to excel
at cricket while at the prestigious British boarding school, Harrow (someone else ran for him), to become a champion swimmer, and according to
British poet Roden Berkeley Wriothesley Noel (1834–1894) in his *Life of
Byron*, to "enter a room quickly, running rather than walking, and stopt
himself by planting his left foot on the ground and resting on it." Neverthe-
less, he seems never to have overcome his shame at being born less than
perfect. On the contrary, he appears to have sought out risky ventures to
prove himself heroic, ending with a trip to Greece that proved fatal.

Nobody doubts that Byron was born with a problem, but accounts disagree as to exactly what was wrong. Was one foot clubbed or were both?
The left or the right? Did Baillie heal only a deformed foot or was it an
entire leg? Inexplicably, the debate continued throughout Byron's life, and
when he died in 1824, of "marsh fever" in Greece where he had gone to

support the campaign for independence from the Turks, the picture was further complicated by conflicting postmortem reports.

Byron's doctor, Julius Michael Millingen (1800–1878), had attended the poet prior to his death; after, he reported that "the only blemish of his body, which might otherwise have vied with that of Apollo himself, was the congenital malformation of his left foot and leg." Byron's intimate Edward John Trelawny (1792–1881) made a hurried trip to Greece to examine the body and noted that "both his feet were clubbed and his legs withered to the knee—the form and features of an Apollo with the feet and legs of a sylvan satyr." Twenty years later, Trelawny changed his mind: "I uncovered the Pilgrim's feet," he wrote, and discovered "the contraction of the back sinews which the doctors call tendon Achilles that prevented his heels touching on the ground and compelled him to walk on the fore part of his feet." Seventy-nine years after that, at a meeting of the Royal Society of Medicine in London in March 1923, Sir Hector Clare Cameron (1843–1928), Dean of Faculties at Glasgow University, displayed models purportedly used to make Byron's boots. They showed two perfect feet. The only possible explanation is that Dr. Cameron, Knight Templar, Commander of the British Empire and holder of a Regius professorship conferred by Queen Victoria, had been snookered.

Meanwhile, Hippocrates' basic stretch-and-bind method with a small change here and there—a boot instead of bindings, a slightly different massage regimen—remained the best known, safest, and most effective approach to clubfoot. However, the notion that if one could cut the constricting tendons and muscles one might release the foot quickly and effectively was beginning to bubble to the surface.

In the first half of the nineteenth century, surgery was a cure worse than the disease. Contemporary statistics suggest that as few as 50 percent of all surgical patients made it out of the hospital alive. The reasons were obvious. Cutting into a living body and keeping that body alive and healthy afterward, creates three challenges: to control bleeding, to subdue pain, and to prevent infection. Ambroise Pare had solved the first in 1552 by teaching surgeons how to tie blood vessels. In 1846, Boston dentist William Morton (1819–1868) demonstrated the anesthetic effectiveness of inhaled ether. Infection remained the final challenge.

In 1816 French surgeon Jacques Mathieu Delpech (1777–1832), Director of the Hôtel-Dieu Saint-Eloi in Montpellier, performed the first modern clubfoot surgery, cutting the Achilles tendon to release the twisted foot. Attempting to reduce the risk of infection by making smaller incisions, Delpech used what he called a "blind" technique on two patients, using a curved knife to make two small cuts to divide the tendon holding the foot in its fixed, clubbed position. Unfortunately, even these smaller surgical wounds tended to become infected, and Delpech abandoned the procedure.

But the experiment obviously made an indelible impression on French novelist Gustave Flaubert (1821–1880), son of the chief surgeon at the Hotel-Dieu hospital in Rouen. In *Madame Bovary*, Flaubert describes in grisly detail the experience of Hippolyte Tautain, the crippled stableman at the Hotel du Lion d'Or in Yonville, the hometown of Emma's cloddish husband, Dr. Charles Bovary.

Charles, Flaubert wrote, had read about a new way to treat clubfoot—it sounds exactly like Delpech's surgery—and decided it would be "a fine thing for Yonville to show how up to date it was by going in for the operative treatment of strephopody." Bullying Hippolyte into the surgery, which turned out to be surprisingly painless, Charles sat back, waiting for applause. Instead, he fell into disaster. The surgical wound became infected; the infection spread. A prominent physician from a neighboring town was called in to amputate Hippolyte's leg, leaving Emma "disgusted with herself for imagining for a single moment that [Charles] could ever do anything well" and poor Charles to feel "that some fatal and incomprehensible influence, he knew not what, was at work around him."

How *could* he have known?

Nine years before Madame Bovary made her entrance onto the French literary stage in 1856,[2] Ignaz Semmelweis (1818–1865) made washing one's hands before examining or treating a patient the rule for students at the maternity clinic at the Vienna General Hospital in Austria, dramatically reducing the incidence of death from puerperal ("childbed") fever. Semmelweis must have been operating on intuition because it was not until 1865, nine years after Flaubert's novel appeared, that Louis Pasteur (1822–1895) theorized that *decay*—infection—was caused by microorganisms carried through the air to land on any available surface, including human skin, mucous membrane . . . or a surgical wound.

Two years later, in 1867, British surgeon Joseph Lister (1785–1869) made the connection between Pasteur's microorganisms and surgical infection. His solution was to bring carbolic acid (phenol) into the operating theater at Glasgow Royal Infirmary. He used the acid to wash his hands and his instruments, saturate the bandages on patients' wounds, and even sprayed it into the air to kill floating germs. This regimen eventually lowered the overall rate of death due to infection after surgery in his hospital from a standard 45 to 50 percent to 15 percent, an astoundingly low figure for the time. The Europeans, particularly the Germans, accepted Lister's antiseptics fairly quickly, but not until 1877, when Lister was named a Professor of Surgery at King's College Hospital in London, did the British medical community fully embrace his advance in surgical patient care.

Of course, hospitals and operating theaters were—and still are—hazardous places, but thanks to Semmelwies, Pasteur, and Lister, the chances of a patient's getting out alive were much improved. So when Delpech put down his scalpel, others were eager to pick up the knife.

Chief among them was Georg Friedrich Louis Stromeyer (or Strohmeyer, 1804—1878), a German pioneer in orthopedic surgery and a founder of *Charité - Universitätsmedizin Berlin*, the medical school for both Humboldt University and the Free University of Berlin. Stromeyer believed that congenital clubfoot was due to a "deficiency" in the internal structure of the malleolus, the bony bump on each side of the ankle; however, he operated mostly on patients whose clubfoot was due to paralysis. His procedure was essentially Delpech's: He inserted a small knife through the skin and underlying tissue to divide the Achilles tendon and free the foot.

One of Stromeyer's patients was the British orthopedic surgeon William John Little (1810–1894), whose foot had been twisted by polio when he was two. Little survived his illness and went on to found the Royal Orthopaedic Hospital in London. He was the first to describe a condition, Little's disease, now known as cerebral palsy. As for his clubfoot, once Stromeyer repaired it in 1836, Little was so grateful that he named his son Louis Stromeyer Little. He also took Stromeyer's procedure back to England, where it was adapted and used to free tendons in other parts of the body as well. In addition, Little, like Hippocrates, emphasized the virtue of manipulating and stretching the tissues around the foot. Unfortunately, like Scarpa, he designed his own punishing device, iron braces extending from the pelvis to the foot.

Dividing the Achilles tendon remained the surgery of choice until midcentury. In 1866, William Adams (1820–1900) was awarded the Royal College of Surgeons' Jacksonian Prize for his essay on the causes and treatment of clubfoot. Adams was the first to examine muscle and bone tissue from a clubfoot under a microscope. His specimens, taken from stillborn children, showed the tissues in the deformed foot to be no different from those in a normal foot, so he opposed dividing the Achilles tendon, at least as a first step in alleviating clubfoot. Like Hippocrates, Pare, and Baillie, Adams believed that what twisted the foot was the force of the surrounding muscles and tendons. He dismissed the Scarpa shoe, but he too had his own mechanical device: a straight splint of sheet metal running the length of the outer leg.

In other words, everything old was new again. Until the end of the nineteenth century when suddenly there really *was* something new: The X-ray, which like so many scientific marvels, was discovered by accident. On November 5, 1895, German physicist Wilhelm Conrad Roentgen (1845–1923) noticed that "rays" (i.e., electron beams) generated by the cathode inside a vacuum tube were not pushed aside by magnetic fields, which meant they could "see" through all kinds of matter. One week later, Dr. Roentgen took an X-ray picture of Mrs. Roentgen's hand: Her wedding ring and her bones were clearly visible along with the pale shadow of her skin.

The new technology fascinated most people, but annoyed others who foresaw a world in which X-rays would let people look through walls and invade privacy, rather like today's airport security devices. Doctors had no such qualms. For them the salient point was that X-rays made it possible to look into the body without cutting and see exactly how bones including those in a clubfoot were deformed or displaced.

Inevitably, this pushed the surgeons treating clubfoot to ever more radical procedures. It became the fashion to divide and lengthen not just the Achilles tendon, but every single tissue around the foot and ankle, or one might remove supposedly deformed bones from the foot. The popularity of both operations faded as it became obvious that one left the patient with a foot that was straight but too stiffened by excessive scar tissue to move freely, and the other gave the patient a foot that was also straight but might never be strong enough to support his or her weight. No wonder that even a medical student like Philip Carey, the long-suffering anti-hero of Somerset Maugham's *Of Human Bondage* (1915), did not seek treatment. Carey's clubfoot, often described by critics as a literary substitute either for Maugham's stutter or for his homosexuality, both of which made him acutely uncomfortable, perfectly captured the idea of a twisted foot as a disabling, but not fatal disorder.

Enter Ignacio Ponseti (1914–2009). Born on the Spanish island of Minorca, Ponseti started his professional life as an orthopedist on the battlefields of the Spanish Civil War, setting fractures and ferrying injured Loyalists across the Pyrenees to France. In 1939, he took himself over the mountains. Two years later he emigrated to Mexico, and from there, a referral from an orthopedist friend sent him to the United States to study and practice orthopedics at the University of Iowa Hospitals and Clinics. Once settled in the American Midwest, the Spanish surgeon discovered large numbers of children whose clubfeet were being treated, but not cured, so he drew up his own stretch-and-bind regimen known as—what else?—the Ponseti method.

Ponseti's method entails manipulating and stretching the tissues to move the foot into a more natural position, beginning as soon as possible, a moment the current Global Help Organization-sponsored Ponseti manual puts at seven to ten days after birth. The regimen begins with the physician's immobilizing the child's foot in a plaster cast for seven days to maintain the correction, then removing the cast, stretching the muscles and tendons again, and replacing the cast. The procedure is repeated each week for a period of three to four months, a regimen that requires as many as twenty different casts per foot. At the end of this part of the treatment, the child is fit with shoes attached to a metal bar to be worn twenty-four hours a day for about six months, and at night for up to two years. Afterward, perhaps three in ten children are said to require a "minor" procedure to lengthen a tendon; fewer than one in one hundred treated with the Ponseti method need major correction surgery.

The introduction of sonography (ultrasound) in the 1950s promised diagnosis as early as eighteen weeks into a pregnancy, making it possible to prepare parents in advance if the image indicated that the child had a clubfoot. As for modern surgery, in 2000, researchers at the Department of Orthopaedic Surgery and Department of Pediatrics at the University of Iowa wrote that "a few reports indicate that surgery [for clubfoot] is almost invariably followed by deep scarring, which appears to be particularly severe in infants. In addition, the average failure rate of clubfoot surgery is 25% [range 13% to 50%] and many complications can occur including wound problems, persistent forefoot supination, loss of reduction and recurrence, overcorrection of the hindfoot, dorsal subluxation of the navicular, and loss of normal motion of the ankle and subtalar joints."

How much better it would be to identify the cause of clubfoot, and thus be able to prevent it.

## Entering the genome

When searching for answers as to how to reduce the risk of a birth defect abnormality such as clubfoot, you can start by asking, "What do the people with this problem have in common?"

Centuries ago the answer was simple: Giving birth to a child with a congenital abnormality was proof of your being out of favor with the gods. Then the blame shifted to human beings, specifically the woman careless enough to think the wrong thoughts or glance the wrong way while pregnant. Simply looking at a rabbit while pregnant might lead to the birth of a baby with a "hare" (cleft) lip, or seeing a one-armed person, a baby with a missing arm. Mary Jane Merrick, mother of John Merrick who was incorrectly diagnosed as suffering from elephantiasis (hence the label "the elephant man"), was said to have been frightened by an elephant at a local fairground.

This theory, known as maternal impression, has not died an easy death. As recently as 2003, Canadian biochemist Ian Pretyman Stevenson (1918–2007), one-time head of the Division of Perceptual (i.e., paranormal) Studies at the University of Virginia, was still trying to prove that a mother's thoughts—or, shades of Bridey Murphy[3]—even her "past lives" could influence fetal development. To this day, a pregnant Parisian, hoping for a boy, may visit the Louvre to stare at paintings of handsome men, most commonly those of the noble persuasion.

When the maternal impression theory did not work out, the search for an explanation turned to more verifiable causes in or out of the womb, such as the physical environment inside the uterus. Perhaps the fetus had been positioned so that its developing foot was bent forward or back under its body. Or crowding in the womb might have twisted the foot out of its natural position or restricted the fetus' movements so that its muscles did not develop the strength required to hold a foot straight. Some studies do show a higher incidence of clubfoot among multiples than among single infants; in the 1960s, researchers at Johns Hopkins proved that paralyzing a chick embryo in the egg prevented its legs from developing normally.

A more demonstrable prenatal culprit is amniotic band syndrome, a condition in which the amnion (the inner layer of the amniotic sac) separates and tears into strings of tissue that may wind around parts of the fetus' body. According to the Fetal Treatment Center at the University of California, San Francisco, this syndrome occurs in as many as one in every 1,200 or as few as one in 15,000 live births. What happens when that is the case depends on where the strings land and how tightly they bind. In the most severe case, amniotic bands wrapped around the head or umbilical cord might be fatal to the fetus. Bands around the face or neck might cause cleft lip and/or palate. Those around fingers or toes might lead to syndactyly (webbing), ordinarily a genetic defect. Strings wrapped tightly enough to cut the blood supply might actually amputate a digit. And a band around the foot might cause a clubfoot.

Late in the nineteenth century, the environment outside the body became suspect. In 1877, Gabriel Madeleine Camille Dareste (1822–1899), professor of zoology at the University of Lille and the first director of the laboratory of teratology at the Ecole des Haute-Etudes (School for advanced Studies in the Social Sciences) in Paris, was awarded the grand prize in physiology by the French Academie des Sciences for his research into how birth defects occur. Exposing chicken embryos to extreme heat, Dareste was able to trigger skeletal and organ abnormalities in the chick developing inside the egg. A century later, his observations on the effects of heat on fetal chickens were extrapolated to human beings, specifically to the effects of temperature on the male reproductive system (the warmer the testes, the lower the production of sperm).

Today, genetic investigators, sensitized by the thalidomide disaster, routinely look for a link between birth defects and known teratogens (from the Greek word *teras* meaning *monster* and *genesis* meaning *creation*)—drugs or chemical agents, including the more than 4,000 chemicals in a single cigarette; some studies, including one from University College London in 2011, suggest that deformed limbs and clubfoot are more common among children born to mothers who smoke while pregnant.

And that brings us to our most essential influence, the place where our lives and all our physicalities originate: our chromosomes and genes.

In 1958, French pediatrician and geneticist Jerome Lejeune (1926–1994) identified a chromosomal anomaly—an extra copy of chromosome 21—as the cause of Down syndrome, the first congenital defect to be conclusively linked to a genetic error. Lejeune's discovery affected both science and culture. It led inevitably to similar discoveries regarding a wide range of congenital abnormalities, which led inevitably to effective prenatal testing. But it also became what you might call the modern version of the Roman Decemvirs' *Table IV*, a decree by which to justify the selective abortion of fetuses with birth defects.

Lejeune, a practicing Catholic, was appalled. In 1972, at a United Nations conference that included delegates who endorsed the idea of selective abortion, Lejeune charged that "here we see an Institute of Health turning itself into an institute of death." That evening, he telephoned his wife to say, "This afternoon I lost my Nobel Prize." He was right. Despite his many honors and the magnitude of his discovery, Lejeune was never nominated for a Nobel. However, on February 19, 2004, at the 10th General Assembly of the Pontifical Academy for Life in Vatican City, Cardinal Fiorenzo Ange-lini, President Emeritus of the Pontifical Council for Health Pastoral Care, announced the beginning of the beatification process of Lejeune, who had been the group's first president. According to his daughter, he would have considered that a more than fair trade.

Long before Gregor Mendel unraveled the mystery of heredity and Lejeune identified the chromosome for Down syndrome, people knew that putting two superior animals together was likely to produce superior off-spring. In *The Republic*, Plato himself proposed using the same method to breed superior people: "You have in your house hunting-dogs and a num-ber of pedigree cocks . . . do not some prove better than the rest? Do you then breed from all indiscriminately, or are you careful to breed from the best? And, again, do you breed from the youngest or the oldest, or, so far as may be, from those in their prime? And if they are not thus bred, you expect, do you not, that your birds and hounds will greatly degenerate? And what of horses and other animals? Is it otherwise with them? . . . How imperative, then, is our need of the highest skill in our rulers, if the prin-ciple holds also for mankind? . . . the best men must cohabit with the best women in as any cases as possible and the worst with the worst in the few-est, and that the offspring of the one must be reared and that of the other not, if the flock is to be as perfect as possible."

And for those who might have missed the message, Plato said it straight out: "The offspring of the inferior, and any of those of the other sort who are born defective, they will properly dispose of in secret, so that no one will know what has become of them. That is the condition of preserving the purity of the guardians' breed."

Plato was not the only one concerned with maintaining the purity of the ruling class; the Spartans' laws regarding infants with birth defects were also aimed at racial purity. By the end of the nineteenth century, the lead-ers of the nascent eugenics movement believed they had discovered a reli-ably scientific justification for selective breeding in Charles Darwin's newly popular theory of natural selection and his open distaste for institutions to protect the disabled. "We civilised [sic] men," Darwin wrote in *The Descent of Man,* "do our utmost to check the process of elimination; we build asy-lums for the imbecile, the maimed and the sick. . . . Thus the weak mem-bers of society propagate their kind."

Soon, this unpleasant view of the disabled made it possible for support-ers of eugenics to establish a firm footing in Britain and the United States. In 1910, Winston Churchill (1874—1965) wrote to then Prime Minister,

Herbert Henry Asquith (1852–1928) that the "unnatural and increasingly rapid growth of the Feeble-Minded and Insane classes, coupled as it is with a steady restriction among all the thrifty, energetic and superior stocks, constitutes a national and race danger which it is impossible to exaggerate. I am convinced that the multiplication of the Feeble-Minded, which is proceeding now at an artificial rate, unchecked by any of the old restraints of nature, and actually fostered by civilised conditions, is a terrible danger to the race." Then Churchill went even farther. He decided that reviving the nineteenth century theory of segregated institutionalization for life was too expensive a solution. Instead, he advocated for forced sterilization, a "simple surgical operation so the inferior could be permitted freely in the world without causing much inconvenience to others."

Ever mindful of their "special relationship" with the British, eugenicists in the United States moved ahead expeditiously to endorse Churchill's "remedy."

In 1907, Indiana became the first state to enact a law that made sterilization mandatory and marriage illegal for mentally handicapped people in state custody. Two years later California followed suit and was soon joined by other states. The Race Betterment Foundation, founded in 1911 in Battle Creek, Michigan, by members of the Kellogg family of cereal fame, underwrote three conferences on "race betterment." The Galton Society, founded in New York City in 1918 to counter the inclusive American Anthropological Association, endorsed the superiority of the Nordic race.

But the intellectual underpinnings of the American campaign came primarily from the Eugenics Records Office (ERO) at Cold Springs Harbor in New York (1910) and the American Eugenics Society (1926), both financed with donations from such philanthropic luminaries as John D. Rockefeller Jr. and the Rockefeller Foundation; Mary Harriman, the widow of railroad tycoon E.H. Harriman; George Eastman of Eastman Kodak; and the Carnegie Institution for Science.

To counter those who questioned the constitutionality of the procedure, the ERO's director, Harry Hamilton Laughlin (1880–1943), wrote a "Model Eugenical Sterilization Law," the basis for a law passed by the Commonwealth of Virginia in 1924. Three years later, in Buck v. Bell, a case involving a seventeen-year-old woman believed at the time to be mentally handicapped, who was the daughter of a mentally handicapped mother and was herself the mother of a mentally handicapped daughter, the U.S. Supreme Court declared the Virginia Eugenical Sterilization Act constitutional. "We have seen more than once that the public welfare may call upon the best citizens for their lives," Oliver Wendell Holmes (1841–1935), whose obviously splendid genes guaranteed him a nicely long life, wrote for the majority: "It would be strange if it could not call upon those who already sap the strength of the State for these lesser sacrifices, often not felt to be such by those concerned, in order to prevent our being swamped with incompetence. It is better for all the world if, instead of waiting to execute degenerate offspring for crime or to let them starve for their imbecility, society can prevent those who are manifestly unfit from continuing their kind. The principle that sustains compulsory vaccination is broad enough to cover cutting the Fallopian tubes. Three generations of imbeciles are enough." Pierce Butler (1866–1939) of South Carolina, the sole dissenting associate justice in the eight-to-one decision, declined to write an opinion.

After that, sterilization programs sprung up like weeds across the country. At the high (or low) point, as many as thirty states had such laws affecting the mentally ill and even persons convicted of such petty offenses as stealing a chicken. Between 1907 and 1981 the surgery was performed on an estimated 65,000 mentally ill Americans; the state of North Carolina alone sterilized more than 7,000 people, most of whom were women whom state and local officials considered to be, for one reason or another, unfit to raise children. The program was officially disbanded in 1977. In 2010, the governor set up an office to locate victims as a prelude to compensation. The North Carolina State House of Representatives endorsed a bill to award $50,000 to persons who had been sterilized. If passed, it would have made North Carolina the first state in the country formally to compensate victims of a eugenics program, but it was killed by the State Senate on June 20, 2011.

Under his own model law, Laughlin, an epileptic, was himself a candidate for forced sterilization. There is no indication that he ever volunteered for the procedure. On the contrary, in 1936, he was awarded an honorary degree from the University of Heidelberg for work described as supportive of the "science of racial cleansing," leading the Nazis to applaud the American eugenicists as allies. Laughlin, whom one biographer described as "among the most racist and anti-Semitic of early twentieth-century eugenicists" does not seem to have objected.

Others did. The German atrocities in the service of racial purity dramatically altered the view of eugenics in the United States, continuing a reversal that began in 1942 when the U.S. Supreme Court overturned an Oklahoma law because it targeted some low-level criminals while excluding the more polite, but equally guilty white collar miscreants: "[S]crutiny of the classification which a State makes in a sterilization law is essential," the Court wrote, "lest unwittingly, or otherwise, invidious discriminations are made against groups or types of individuals in violation of the constitutional guaranty of just and equal laws." Although the Nixon administration broadly increased funding for the voluntary sterilization of poor Americans through the Medicaid program (many believe that the majority of these operations were involuntary), there was no turning back. In 1981, the state of Oregon performed the last legal forced sterilization in the United States.

The conundrum of selective abortion, however, is still with us, as is the question of what constitutes an abnormality incompatible with an acceptable life.

In Great Britain, late abortion (after twenty-four weeks), is currently legal only in cases of "serious risk to life of the woman or severe fetal abnormality." Church of England cleric Joanna Jepson, who was successfully treated in her late teens for a congenital anomaly that set her upper jaw so far out in front of her lower jaw that her teeth could not meet, has long campaigned against aborting fetuses with cleft palate. In 2003, she went to the High Court of Justice, which hears important civil cases in Britain, to argue that terminations based on a clubfoot were illegal was not a "serious disability." Three years later, when the *The Sunday Times* of London published figures from the British Office for National Statistics showing that more than twenty women had obtained late-term abortions after scans revealed they were carrying a fetus with a clubfoot. Jepson announced herself appalled that women were "under pressure to abort . . . in a situation where there is a treatable problem such as clubfoot."

The issue of acceptable disability remains—to say the least—contentious. But Lejeune's discovery of Trisomy 21 dramatically quickened and expanded the search for problematic genes, sometimes with positive results for the human foot.

Geneticists now know, for example, that polydactyly (extra toes and/ or fingers) is carried as an autosomal dominant trait, an effect that occurs even if the fetus inherits a problematic gene only from one parent (a recessive trait requires a copy of the gene from both parents). Webbing of the toes or fingers, failure of the digits to separate normally around the sixth to eighth week of pregnancy, is also an autosomal dominant trait.

As for clubfoot, even before there were measured studies, people understood that clubfoot, like some other abnormalities, might "run in families." All you had to do was look at your friends and neighbors to see that if one member of a clan had a clubfoot, another family member somewhere in the generations, might also have one. But scientists prefer to nail down their theories with numbers; in 1912, Alfred Ehrenfried (1880–1951) of Children's Hospital in Boston produced the first set for clubfoot. Toting up five years' worth of statistics on 232 young patients in the hospital's orthopedic outpatient department and excluding those whose clubfoot occurred after paralysis or an injury, Ehrenfried was able to identify a family relationship for nearly one in seven cases, an incidence of 14 percent. Two years later, when the number of patients had risen to 342, the incidence of a family history among patients with *talipes equinus varus* (the foot turned down and in toward the other foot) was 100 percent.

Today it is understood that about one in four people with clubfoot have a relative with the same deformity. Parents who have already given birth to a child with a clubfoot have a 10 percent risk of giving birth to a second child with the deformity. If one monozygotic (*from one egg,* i.e., identical) twin has a clubfoot, the odds of the other twin's having the problem is about one in three versus one in thirty-three for nonidentical twins. Although Ehrenfried found no clubfeet among his "colored" patients, modern population studies show that the incidence of clubfoot differs markedly among ethnic groups. Clubfoot is most common among Polynesians (Hawaiians and Maoris) for whom the incidence is seven per 1,000 live births, and least likely among the Chinese for whom the incidence is about one for every 3,000 live births.

Regardless of race or ethnicity, clubfoot is twice as common among boys as among girls. About 20 percent of all cases of clubfoot are associated with other genetic defects such as distal arthrogryposis (contracted joints that make it difficult to move the hands and feet), congenital myotonic dystrophy (progressive muscle wasting), spina bifida, and Trisomy 18 (an extra copy of the chromosome that is linked to low birth weight, a small head with a small jaw and low-set ears, mental retardation, clenched hand, and undeveloped fingernails).

Clearly, something in our DNA has gone amiss here.

But what? Which gene is to blame?

A normal human cell has forty-six chromosomes. Half of the chromosomes are provided by the mother via her egg and half by the father via his sperm. The chromosomes occur in pairs, so naming them is easy: All you need is the ability to count to twenty-three. Genes are an exponentially different matter. Together, the twenty-three human chromosomes hold as many as 25,000 genes, each with its special individual role to play in the development of mind and body. To name them, you need to know what they do. For example, *Pitx1* is the abbreviation for a gene whose official name is *paired-like homeodomain 1*. This gene contains a sequence (homeodomain) of DNA that affects the development of a particular part or parts of the body. The *Pitx1* gene, located on human chromosome 5, is also found in other animals as well as plants and fungi. In each species, the gene regulates patterns of anatomic development, including the symmetry of the left and right sides of the organisms. It is not easy to spot an asymmetry caused by a problem with the *Pitx1* gene in mushrooms or daffodils, but if something goes wrong with a vertebrate's *Pitx1, the* results are likely to be painfully obvious.

In the mid-1990s, geneticists at the Howard Hughes Medical Institute in California identified a link between *Pitx1* and hind limb development in mice. Repeated studies show that mutated or missing *Pixt1* genes result in shorter hind leg bones and missing toes in the rodents, deformed wings in chickens, and pelvic abnormalities in manatees and sticklebacks (also known as sticklefish, varieties of small fresh- and seawater fish distinguished by spines along the top).

In 2008, researchers at the Washington University School of Medicine in St. Louis found evidence of a similar genetic link in human beings. For us, *Pitx1*, which activates *TBX4*, a gene on chromosome 17 that regulates the acquisition of two hind limbs, is vital to the development of a normal leg.

The Washington University team's report in *The American Journal of Human Genetics* was based on their study of a *Pitx1* mutation among members of an Iowa family with a history of malformations of the lower leg. In this one large family, at least eighteen of thirty-five relatives had one or more specific limb deficiencies or deformities such as a missing lower leg bone (tibia), extra toes, and/or a clubfoot. Interestingly, five women in the clan who carried the gene did not develop any deformities, which suggests a male susceptibility. Two years later, the same Washington University team defined irregularities on chromosome 17 in people with isolated clubfoot (not associated with other abnormalities). In 2011, they were able to confirm the link between a mutation in one region of the *Pitx1* gene and clubfoot. "Much remains to be identified regarding both the genetic and mechanistic aspects of this condition," they said, but "this discovery opens up the possibility of clinical genetic testing for patients with familial isolated clubfoot and provides important insight into the developmental pathway responsible for human limb birth defects."

But not solutions for fixing them.

In 1981, University of California pediatric surgeon Michael Harrison, now the Director Emeritus, Fetal Treatment Center, Division of Pediatric Surgery, and Professor Emeritus of Surgery, Pediatrics, Obstetrics, Gynecology and Reproductive Sciences at the University of California at San Francisco (UCSF), performed the first open surgery on a fetus. His book, *The Unborn Patient: Prenatal Diagnosis and Treatment* (1984), literally created the specialty of fetal surgery. Today, the technique is an accepted treatment for a handful of congenital abnormalities such as a malformed heart, tracheal atresia (an abnormal windpipe), a narrowing or obstruction of the urinary tract, a pulmonary lesion such as a cyst that makes breathing difficult, a disorder of the spinal cord such as spina bifida, or a cleft lip and/or palate, all problems that modern technology may diagnose before the baby is born.

A clubfoot is not on this fearsome list. A foot, after all, is only a foot; even when bent it is serviceable. But a clubfoot is a deformity, and all deformities have effects—although not necessarily the ones you might expect.

Health care professionals are sometimes surprised that parents of children with what is considered a minor problem, such as a clubfoot, evince the same shock, anger, guilt, and depression as parents whose children are born with far more debilitating abnormalities. A team of geneticists at the University of Aberdeen, Royal Children's Hospital in Scotland and Bournemouth University in England says the experience is real. In 2011, the doctors interviewed fifteen families of children born with a clubfoot. Their report in *The International Journal of Orthopaedic and Trauma Nursing* captured the fears of parents who face financial and psychological problems along with the worry that their children will suffer from being "different."

"In the 15 years I have worked with families of children with clubfoot," study leader Zosia Miedzybrodzka wrote, "I have become aware that the condition is more of an issue for families than healthcare professionals believe it to be. The generally held view is that because the condition is treatable it does not affect families too much. However our study shows that this is not the case. The treatment for clubfoot puts a huge burden on families who have to deal with months and years of treatment with plaster casts and then boots with bars on their child's legs, as well as frequent visits to the hospital."

The children, it seems, may see things differently. When Edwin van Teijlingen of Bournemouth University interviewed young patients for the Aberdeen study, he discovered that they did not necessarily share their parents' concerns, at least not to the same extent. True, the children knew they had a problem, but did not see themselves as being unlike others their own age.

This optimism is likely conditioned by the availability of treatment, but there is another possibility as well. Nothing so strengthens trust in your society as a law that says, "You are one of us." For these children, as for others with far more disabling conditions, the world is larger than it once was and their access to it is guaranteed by twenty-five years of legislation to ensure protection against discrimination: The Americans with Disabilities Act of 1990 (amended and updated in 2004), the British Disability Discrimination Act 1995 (updated and replaced by the Equality Act 2010), the Canadian Employment Equity Act 1995, and similar though sometimes less-expansive legislation in Europe and the Middle East. Our view of disability and the disabled has been irreversibly altered by these laws spurred not only by the number of infants born with birth defects, but also by growing numbers of those disabled by war and disease who, thanks to dramatic advances in medicine, survive to cope with the results of disabling injury and loss.

As noted earlier, in the ancient world, infants born with a clubfoot were spared the fate of other children with birth defects. Tut's disability was royal and thus seemingly acceptable. The Greeks and Romans raised acceptance to a higher level. Hephestus (Vulcan to the Romans), was the son of Zeus and in his own right the god of fire and the smith who crafted Hermes' winged helmet and sandals. More to the point, Hephestus was born lame, with one or possibly two clubfeet, thus making the deformity not only royal but certifiably divine. In Europe, as late as the Middle Ages, a clubfoot was regarded by some as a sign of the Devil, but by the eighteenth century, even moderate treatment allowed clubfooted Byron and Talleyrand to function well in societies that prized their intellectual talents above their physical deformities. Today, when treatment begins soon after birth, those born with a clubfoot are likely to walk, skate, or run past their disability to a thoroughly normal life with impressive regularity.[4]

In the United States alone, the list of well-known people born with clubfoot includes Thaddeus Stevens (1792–1868) and DeWitt Clinton Fort (1830–1868). Neither Stevens, an abolitionist radical Republican who defended runaway slaves and helped draft the Fourteenth Amendment to the U.S. Constitution, nor Fort, a member of the 2nd Missouri Cavalry most famous for escaping from a Northern prison, carrying with him the oversize extra-long, heavy double-barreled shotgun he hauled along with him from Texas on his way to join the Confederate forces, was ever slowed by his disabled foot.

In the Arts, there's Gary Burghoff, "Radar" O'Reilly in the *M\*A\*S\*H* television series who was also born with shorter-than-normal fingers on one hand, a deformity he often camouflaged by carrying a clipboard on screen; filmmaker David Lynch, best known for the movie *Blue Velvet* and what may be the first intentionally weird television series, *Twin Peaks*; and the late more-or-less honorary Yank, British actor-comedian-musician Dudley Moore (1935–2002) who wore one shoe with a built-up sole to compensate for his clubfoot.

Among American male athletes, count Hall of Famer Troy Aikman, former quarterback for the Dallas Cowboys; Red Sox and Pirates infielder Freddy Sanchez; New England Patriots offensive lineman Brian Simmons; pitchers Larry Sherry (named most valuable player in the 1959 World Series when the Dodgers took their first championship after moving from Brooklyn); and Jim Mier, retired from the Marlins in 2003.

Finally, 2,500 years or so after the Greek poet philosopher Hermodorus walked out of the shadow of Mt. Olympus to Rome—unfortunately to assist in demonizing those whom Plato called "the other sort who are born defective,"—Kristi Yamaguchi, the 1992 Olympic singles women's figure skating champion, and Mia Hamm, a member of the American women's champion soccer team at the 1996 Olympics both had the birth defect.

It is pleasant to imagine these women speeding to the gold on winged sneakers or skates forged, at least in spirit, by a true Olympian, Hephestus, the clubfooted son of Zeus.

## Notes

(1) The name Scarpa is still linked to the word "shoe," though not the way Scarpa intended. Today, SCARPA is an acronym for *Società Calzaturiera Asolana Riunita Pedemontana Anonima* (Associated Shoe Manufacturing Company of the Asolo Mountain Area), now owned by the Parisotto family of Asolo in northern Italy. The first Parisotto, Luigi, went to work for SCARPA at thirteen in 1942; fourteen years later he and his brothers bought the company that today is best known for its innovations in outdoor gear. This includes the first high-altitude plastic boot, the first Gore-Tex boot, and the first plastic Tele-markSki skiing boot, designed, of course, to keep the foot safe and steady. To which Pare might well have said, *"Plus ça change, plus c'est la meme chose"* ("The more things change, the more they stay the same.")

(2) The first English edition of *Madame Bovary* appeared in 1886, thirty years after its debut in France and three years after Flaubert's death. It was translated by Eleanor "Tussy" Marx Aveling (1855–1898), one of six children of Karl Marx, only three of whom survived into adulthood. Like Emma Bovary, who unsuccessfully rebelled against the Bourgeois society, Eleanor died a suicide, of poison. In one of those "we all know each other" moments, the doctor

who pronounced her dead was Henry Shakleton, father of the future polar explorer, Henry Ernest Shakleton.

(3) In 1956, hypnotist Morey Bernstein published *The Search for Bridey Murphy,* the story of Virginia Tighe, a Colorado housewife who, under hypnosis, recounted her "past life" as Bridey Murphy of Cork, Ireland. Based on Tighe's knowledge of Cork and its inhabitants, plus her lilting Irish brogue, Bernstein insisted that his heroine's story was authentic. *Life Magazine*'s March 19, 1956, story "Bridey Murphy Puts Nation Into a Hypnotizzy" made Virginia/Bridey a national phenomenon, but as details of Tighe's life emerged—one of her childhood neighbors was named Bridey Murphy Corkell; there was no birth certificate for Bridey in Cork and no death certificate in Belfast where Tighe said she had died—the story becme less interesting and faded from the headlines. Tighe died on July 13, 1995.

(4) Today, every year around the world an estimated 135,000 to 150,000 infants are born with a clubfoot. The following table shows the estimated numbers for selected countries. The actual numbers may be higher because these figures, based on one case per 1,000 live births in the population, are drawn from the Global Clubfoot Initiative's record of the number of persons in the selected country currently being treated with the Ponseti method.

| Country | Estimated number of people with clubfoot* | Number being treated each year[†] |
|---------|-------------------------------------------|-----------------------------------|
| Bangladesh | 141,340 | 1,351 (2010) |
| Democratic Republic of the Congo | 61,315 | 554 |
| Ethiopia | 71,336 | 651 |
| Ghana | 20,757 | 282 |
| Kenya | 32,982 | 691 |
| Laos | 6,068 | 78 |
| Niger | 11,360 | 17 |
| Paraguay | 6,191 | 56 |
| Tanzania | 36,070 | 75 |
| Zambia | 11,025 | 548 |

* Adapted from Cure Research.com, http://www.cureresearch.com/c/clubfoot/stats-country_printer.htm

[†] Global Clubfoot Initiative, http://globalclubfoot.org/world-data. Except where noted, the figures are for 2009.

# 3

# DIFFERENCE

"If a man be courteous to a stranger,
it shows he is a citizen of the world."

Francis Bacon, *Essays* (1625)

WHETHER FIERCE AS A LION or meek as a mouse, all human beings and other domesticated animals are born with two instinctive fears: the fear of falling and the fear of loud noises.

Time adds two more terrors: the fear of separation and the fear of strangers.

Early on, Freud blamed separation anxiety on the trauma of birth, the primary separation, and then on the absence of warmth from the nursing mother. British psychiatrist John Bowlby (1907–1990) re-cast it as a mistake in attachment behavior, the socialization that begins with the newborn's relationship to her mother and then, with that secure, spreads out to encompass other people and experiences.

Bowlby, who codified the stages of attachment behavior, was raised by parents who thought that too much attention and affection would spoil a child; as an adult he went full steam ahead in the opposite direction. After graduating from Trinity College, Cambridge, he served in the Royal Army Medical Corps during World War II; was then named director of London's leading psychiatric facility, the Tavistock Clinic; and eventually became a mental health consultant to the World Health Organization, which in 1949 asked him to evaluate the mental health of displaced and homeless children in Europe. His report, *Maternal Care and Mental Health* (1951), proposed that a child's attachment to a stable nurturing figure was in fact an evolutionary imperative built into the infant's brain to benefit her mental health for the rest of her life. He was right. As anyone who has seen one of those heartbreaking National Geographic documentaries about an orphaned baby lion or elephant knows, throughout the animal kingdom infants who are protected from the day of their birth are more likely to survive and thrive.

Like Bowlby, Harry Harlow (1905–1981) was raised by a chilly mother. Born Harry Israel, he took his father's middle name at the request of his mentor at Stanford University, Lewis Terman (1877–1956), the American psychologist best known for developing IQ tests. Despite his name, Harlow was not Jewish, and Terman thought he would do better with a name changed to avoid the anti-Semitism then rampant in the science research community.

Harlow challenged Freud's theory that feeding was the essential component in the mother/child bond. To prove his point, he separated newborn Rhesus monkeys from their mothers, putting the infants in cages with two substitutes: One was covered with soft cloth, but provided no food; the other was a bare naked wire form equipped with a formula-filled baby bottle. Given the choice, the babies spent more time with the soft substitute than with the nursing one, proving that Freud had missed the more important point, Harlow's conclusion that "contact comfort is a variable of overwhelming importance in the development of affectional response, whereas lactation is a variable of negligible importance."

Freud eventually pinned his own phobias not on the moment of birth, but on the sudden disappearance of his beloved nanny when he was three; the woman was arrested for stealing from the Freud household, an event he called the "primary originator" of his own neurosis. As for Harlow's baby monkeys, they were so traumatized by separation from their mothers that they never recovered. As adults, they found mating difficult and could not properly care for their own infants. Harlow's work, still considered essential to an understanding of the foundation of human love, earned him a National Medal of Science (1967) and a Gold Medal from the American Psychological Foundation (1973). But the experiment, widely condemned as extraordinarily cruel, became a rallying cry for the Animal Rights Movement in the United States.

As for the second social anxiety, the fear of strangers, no one yet seems to have pinpointed the exact moment when (or whether) stranger anxiety happens to puppies and kittens and monkeys, but for human babies Bowlby put it at some time between eight and twelve months of age when infants begin to shy away from new and unfamiliar faces. Most of us eventually learn to make new friends, or at least to mask our unease at the presence of new people, but as a society our infantile fear is ingrained as xenophobia, a generalized dread of the Others, those people outside the clan from whom we must protect ourselves.

To do that, we need to name the things that make *us* different from *them* and then erect a protective barrier.

One historical example is the inability to pronounce "sh," a failing used to an advantage by the men of Gilead, who after beating the Ephraimites in battle (Judges 12:5–6), took up posts along the banks of the River Jordan to keep their foes from crossing over on their way home. Simply by asking travelers to say *shibboleth*, which the Ephraimites pronounced *sibboleth*, the good folk of Gilead were able efficiently to identify and eliminate as many as 42,000 of their enemies.

There are, of course, other tests for Other-ness: skin color perhaps, curly (or straight) hair, the shape of the nose . . . or the height of the arch in the foot, a feature clearly visible and distinct in the left foot of Leonardo's *Vitruvian Man*.

## Creating the curve

Our feet differ from those of our primate cousins in two important details. The first is the position of the big toe, close to the four others so that our feet no longer function as hands. The second is the arch at midfoot.

Most vertebrates—think dogs, cats, lizards, elephants—have feet that sit flat on the ground. Primates from monkeys to apes and the prosimian aye-ayes, indris, lemurs, lorises, monkeys, pottos, and tarsiers (whose elongated ankle bones allow the six-inch-long creatures to propel themselves into leaps as long as nine feet) have flat but flexible feet with those grasping opposable thumb-like big toes.

Our feet are neither completely flat nor completely flexible. Rather they are stabilized by arches built of bones wrapped in ligaments, tendons, and muscles strong enough to support our weight when we stand up. Our arches seem to have been with us since we stepped on to the path leading to human. In 2011, nearly forty years after our famous fossilized *Australopithecus afarensis* ancestor Lucy was uncovered in Ethiopia, an international team of archeologists returned to the scene and discovered a metatarsal bone that had somehow been missed the first time around. This small bone, which links a toe to the rest of the foot, proved that sometime between 3.9 and 2.9 million years ago, Lucy walked upright on arched feet.

Nobody knows exactly how many arches there were in Lucy's feet. Ours have three: the medial longitudinal arch, the lateral longitudinal arch, and the transverse arch.

The medial longitudinal arch, which runs along the inside edge of your foot front to back from toe to heel, comprises the calcaneus (heel bone), talus (ankle bone), navicular (the bone that connects ankle and heel), cuneiform and three metatarsal bones. The bones are supported by the plantar calcaneonavicular and the deltoid ligaments, the tibialis posterior and anterior tendons, the peroneus longus muscle coming down from the tibia (leg bone), and the muscles on the bottom of the foot.

The lateral longitudinal arch runs along the outside of your foot, again front to back, toe to heel. If you did not even know you had this arch, that's not surprising. It is hard to see except in people who have extremely high natural arches. This arch is very sturdy, with limited flexibility among its bones (the *calcaneus, cuboid* [*tarsal*], fourth and fifth *metatarsals*), tendons (the long plantar along the bottom of the foot to connect the cuboid and calcaneus bones and the extensor tendon that lifts the toes), and the muscles of the fifth toe.

As its name implies, the transverse arch crosses your foot from one side to the other, about halfway between your toes and your heel. This arch is supported by the interosseous tendons attached to the bones in midfoot bones, the plantar and dorsal ligaments on the sole of the foot, the muscles of the first and fifth toes, and the peroneus longus, whose tendons crisscross all the arches.

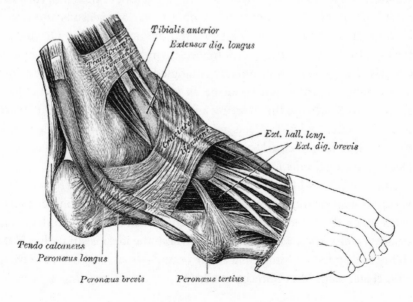

Like every other part of the human body, the arched foot comes with its own standards of grace and beauty, duly immortalized by sculptors and painters such as Sandro Botticelli (c. 1445 –1510) and Edgar Degas (1834–1917).

Standing in front of Botticelli's *The Birth of Venus,* most of us focus quite naturally on the lush flesh tones (below, in shades of gray, alas) of

the goddess rising from her scallop shell. To see the ideally arched foot, you have to ignore her and zero in on the right foot of the Hora (one of a coterie of mythical creatures who symbolize various seasons and times) holding the cloak in which she will wrap the naked Venus. Or you can look left to Zephyrus, the god of the west wind, and his wife, the nymph Chloris. The two are so intimately entangled that trying to figure which foot belongs to what figure can be as puzzling as deciding whether a staircase in *Relativity,* the dizzying anti-gravity lithograph by Dutch artist M.C. Escher (1898–1972), is going up or coming down. But Zephyrus' (or Chloris') foot is clearly arched. So are those of the ballerinas in the Degas painting, *In the Dance Studio* (facing).

In the ballet world, this high-arch extremity is known as a *banana foot.* When a ballerina lucky enough to have banana feet stands on her toes in ballet shoes, the curved feet in the similarly curved toe shoes produce classically graceful lines. Male dancers, for whom the ideal is strength and the ability to leap tall buildings in a single bound—or at least from the wings onto center stage like Albrecht in *Giselle*—can flex a foot into a curve for

effect, but they dance more or less flat on their feet. Although this jars the foot and ankle, it may spare the men crooked toes, darkened nails, calluses, corns, and bunions, all common among the ballerinas. Unless the men are members of *Les Ballets Trocadero de Monte Carlo*, the fearless group of totally serious, but utterly funny dance parodists who first stood up in size 12 toe shoes in New York City's Off-Off Broadway theaters and lofts in 1974. Since then, the "Trocs" have been en pointe on stage to nonstop applause all around the world; of course they are also at risk for crooked toes, darkened nails, calluses, corns, and bunions.

Is there an alternative? Well, they could switch to flatfoot dancing, named not for the foot, but for the way the foot moves. Flatfoot dancing is a step dance, a choreography in which the stamping, brushing, or tapping movement of the foot against the floor is used to produce rhythmic sounds in time with the tune, usually played on a fiddle. The dance, also known as Appalachian clogging, is similar to the Irish clog, not surprising given the Scotch–Irish ancestry of many of the original settlers in the North Carolina hills. The American version is more irregular than the Irish and more connected to the floor, the theory being that if your audience can see the soles of your feet, that's not flatfoot dancing. And the New World clog is most commonly an individual performance rather than an ensemble such as *Riverdance, Lord of the Dance,* and *Celtic Tiger,* the clog-dance spectacles created, through odd coincidence or a really great cosmic joke, by an Irish-American dancer from Chicago named Michael *Flat*ley.

Meanwhile, back at the classic barre, you can sometimes find people with flatfeet sent there specifically to work their feet into arches. True, ballet can strengthen the muscles in the foot. And sometimes just being there yields unexpected benefits. As a girl, actress Jane Seymour had flatfeet and a speech impediment. "So I was immediately enrolled in ballet lessons to get rid of the flatfeet," she says on her website, "and enrolled in speech classes, which of course ended up making me love the theatre."

But newly banana-ed feet?

Alas, no. True arches are born, not made. They come with our genes.

Virtually all human babies arrive with flatfeet. At birth, the tendons that hold the joints and bone in the foot together are loose, tightening only as an infant grows and the arch begins to form between twelve and eighteen months of age. By age three, the arch is usually firmly in place. Unless it isn't: As adults, approximately 80 percent of us have arches; the other 20 percent don't.

Some people have flatfeet due to a congenital condition such as cerebral palsy which weakens muscles and tendons, including those in the arch. Others acquire their flatfeet later in life through normal aging, or after an injury that tears the posterior tibial tendon (the tissue that runs down the sole of the foot to hold the bones forming the arch firm) or an illness. This circumstance, quite reasonably labeled *adult (acquired) flatfoot,* is four times more common among older women (age fifty and up) than among men of the same age. Depending on the cause, the newly fallen arch may be either of no concern or serious enough to require a surgical fix.

The most common flatfoot is a flexible one. Flexible flatfeet form an arch when you stand on your toes, but the arch disappears as soon as you stand down again on the whole foot. This type of flatfoot is often due to benign hypermobility joint syndrome (BHJS), a condition familiarly known as being "double jointed." BHJS may run in families, perhaps along with curvature of the spine; a dislocated hip, elbow, knee, or shoulder; and/or frequent ankle or wrist sprains. Up to 40 percent of school age children, more frequently girls than boys, have BHJS bendable joints that allow them to fold a thumb down toward the arm, sometimes all the way down, or bend the other fingers (especially the pinky) back to touch the arm. Although that may make a child awesome in the schoolyard or on the gymnastics team, it may also lead to complaints of discomfort after activity, more commonly among younger children. By the teen years, most muscles and joints tend to tighten naturally, although children with BHJS who have flatfeet tend to have the same flatfeet as they grow older. But Nature may compensate: Sometimes people whose joints were hypermobile in childhood have fewer problems and better foot function in older age than those whose joints were always normally mobile.

A foot that doesn't arch when its owner stands on tiptoe is called a rigid flatfoot. Rigid flatness, which is less common than flexible flatness, is likely due to tarsal coalition, a condition in which the bones at the back of the foot are cemented together to create an abnormal connection between either the calcaneus (heel) and tarsal bones or the calcaneus and navicular bones, or less commonly, both the tarsal and the navicular bones. Experts at Boston Children's Hospital say that a tarsal coalition may be "a genetic error in the dividing of embryonic cells that form the tarsal bones during fetal development" or that it may be "triggered by: trauma to the area, infection, self-fusion of a joint caused by advanced arthritis (rare in children)."

So what's a healthy arch? How high is *high*, and how flat is *flat*?

In 2003, the National Institutes of Health funded the study Prevalence of Foot and Ankle Conditions in a Multiethnic Community Sample of Older adults. The researchers from the New England Research Institutes, Boston University' School of Medicine, Sargent College, and the New England Baptist Hospital defined a high arch as one under which one of them could put two fingers when the person was standing. Of course, because fingers differ in width, this begs the question, "Whose fingers?"

Another similarly "iffy" test asks you to stand a pair of your shoes on a flat surface, the left shoe on the left, the right shoe on the right, and see if they lean in toward each other. If they do, that suggests your feet are also leaning inward because your arches are not arched.

And let us not forget the Wet Foot Test. For this one, you dunk your feet in water, stand on a piece of paper, step off, and look at the impression you left behind. Is there a blank space curving in from the pad behind your big toe toward your heel where your arch should lift your foot up off the paper? You have adequate arches. No blank space? Flatfeet, like the impression left by Robinson Crusoe's accidental companion, Friday. "It happened one day, about noon," Daniel Defoe's Crusoe recalls, "going towards my boat, I was exceedingly surprised with the print of a man's naked foot on the shore, which was very plain to be seen in the sand. I stood like one thunderstruck, or as if I had seen an apparition. I listened, I looked around me, I could hear nothing, nor see anything. I went up to a rising ground to look farther.

I went up the shore, and down the shore, but it was all one; I could see no other impression but that one. I went to it again to see if there were any more, and to observe if it might not be my fancy; but there was no room for that, for there was exactly the *very print of a foot - toes, heel, and every part of a foot* (emphasis added)." No empty space. No arch.

On the other hand, or foot, if the impression you leave behind shows only the heel and toes, that's a foot with a high arch, known medically as pes cavus—from the Latin words *pes* meaning *foot* and cavus meaning *hollow*, as in the hollow space under the high arch. Would you be surprised to learn that like flat arches, high ones can also be annoying? Normal arches flatten slightly when a person is standing; high arches hold their distinct shape, robbing the foot of some of its weight-bearing potential and stressing the bones in the heel and the front of the foot. This discomfort may travel up the leg to the ankle, knee, and hip joints. Sometimes a person with pes cavus "clutches" at the floor with his toes; persistent clutching may lead to arched toes with calluses on top. Like flatfeet, high arches may run in families and are normally of no serious consequence. But if the arching occurs suddenly and affects only one foot, it may be a sign of a neurological disorder such as Charcot-Marie-Tooth disease (a form of muscular dystrophy in which victims slowly lose the use of their arms and legs), Friedreich's ataxia (a form of progressive nerve degeneration), or any one of several hereditary forms of progressive muscle and/or nerve damage, each of which requires a doctor's care.

Regardless of the arch, the soles of our feet are crime-fighting tools that can testify to our individual identity.

Our mammalian skin has two top layers: the outermost epidermis and the dermis underneath. Under the dermis is a layer of fatty tissue containing sweat glands that secrete salty watery fluid and sebaceous glands that secrete oils to lubricate and protect the epidermis. When you touch something, these liquids leave behind an impression of the pattern formed by the ridges and furrows of the epidermis. This is the basis for the statement that "every contact leaves a trace," a forensic dictum known as Locard's Exchange Principle, named for Edmund Locard (1877–1966), director of the world's very first crime laboratory, established in Lyons, France, in 1910 (the first such laboratory in the United States opened in Los Angeles in 1923). In 1920, Locard published *L'enquete criminelle et les methodes scientifiques* (*The Criminal Investigation and the Scientific Method*), a book in which he asserts that "wherever he [the criminal] steps, whatever he touches, whatever he leaves will serve as a silent witness against him."

That "witness" is wholly individual. No two human beings, not even identical twins, have exactly the same DNA or exactly the same foot- or hand- or fingerprint. The ridges, furrows, and minutia points (marks at the end of a ridge) on the surface of the skin that comprise fingerprint development in the womb, probably between the third and fifth month of fetal development, are with you forever. Many malefactors have tried to change their prints by burning or cutting their fingertips. It's pretty much a waste of time. Slice the skin off the entire fingertip, and your body rushes in to repair the damage, the skin heals, and the patterns engraved during fetal life emerge once more. Cutting randomly into the tip may cause scarring that alters the basic print, but leaves either a recognizable partial print or creates a new print unique to you. Either way, the prints remain in place even after death when they continue to provide identification until the flesh itself decomposes and melts away.

For a while in the 1990s, many American hospitals foot-printed newborns, ostensibly to prevent accidental mix-ups. The prints make nice souvenirs, but according to the FBI Criminal Justice Services Division, although the ridges that make up the newborn's prints are fully developed, the feet, palms, and fingers are so small and the ridges so close together that the prints turn up as "blobs of ink" rather than distinct impressions. To get a print clear enough to serve as identification you must wait until the child is about five years old. By that time, the soles of the feet as well as the palms and the fingertips are large enough and the patterns of ridges and furrows sufficiently pronounced to make a useful impression. This may be one reason why The Infant Protection and Baby Switching Prevention Act—which includes footprinting among the security procedures to reduce the likelihood of infant patient abduction and baby switching and has been proposed in the U.S. House of Representatives every year since 1999—has never moved out of committee to a vote by the full House.

Adult footprints, like adult fingerprints, are fully formed. In 2011, scientists at Shinshu University in Tokida, Japan, used three-dimensional image-processing techniques to analyze the patterns of prints left by the footsteps of 104 volunteers when the heel hits, the foot rolls toward the toes, and the toes push off for the next step. The results were marks so distinct to an individual that footprints may rank alongside voice recognition, retinal scans, and of course fingerprints, as a tool for forensic identification. Daniel Defoe (1660–1731) wrote *Robinson Crusoe* in 1719, more than a century before it was discovered that the footprint—not Friday's plain flat impression on the sand, but a clear impression of lines on paper or some other background—is a characteristic as individual as our DNA. British research podiatrist Wesley Vernon, co-author of *Forensic Podiatry, Principles and Methods* (New York: Humana Press, 2011) considers the footprint scene the first example of forensic podiatry.

For the moment, footprinted or not, most congenital flatfeet present no serious medical problem for either children or adults.

But medicine is not sociology, nor is it religion.

And therein trouble lurks.

## The discriminating drama of the different foot

If our own culture had evolved under different circumstances, say on the African continent rather than in Western Europe, our standards of Otherness might be tall Eastern African versus broader more muscular Western African versus the much shorter Central African pygmy and various shades of black, brown, and tan with the norm and the Other depending, as usual, on who came in first in the wars, cultural, and otherwise—that is, the ones with weapons. But Westerners are, for good or ill, mostly the heirs of those dreaded "Dead White Males," European men who over the centuries wrote our literature and our music, created our pictures and statues, and created the laws by which we live. As a result, our differences, for good or ill, have usually been those things that separate the Western European from the rest of the world.

The exception is women, proving that you don't have to be in the minority to be an outsider. From the beginning, although they have always represented at least half and sometimes more than half the population, women have consistently been defined by the men who wrote the rule books as "not us." Or as Aristotle put it: "The female is a female by virtue of a certain lack of qualities; we should regard the female nature as afflicted with a natural defectiveness." For Aristotle the moralist, the "defects" included the "facts" that women had fewer teeth than men and that only fair-skinned women experienced orgasm. Shame on him. Even worse, the quote does not appear in *Bartlett's Familiar Quotations*. Shame on them.

As for men, the *Vitruvian Man's* is not the only ideal body. The ideal Buddhist body is described in the *Pali Canon*, the first written version of the teachings of Siddhartha Gautama (c. 563–4 to 483 [or 411 or 400]). The *Canon* is one version of the *Tripitaka* (Sanskrit for three baskets) comprising the *Vinaya* that teaches conduct, the *Sutra* that concerns meditation, and the *Abhidharma* that deals with the totality of human knowledge. Considered too sacred to be committed to writing, these words were not written down until the first century. They were kept alive by monks who for five centuries memorized and chanted them to the next generation. The *Canon* lists thirty-two characteristics of a Great Man including his foot, which is equal in length to its partner with a sole engraved with netlike lines that make it extraordinarily flexible (ordinary human beings with only two or three of these lines can just move their feet up and down), tube-shape toes all equal in length, smooth toenails that turn slight up at the tip, oval (not round) heel. And of course, a distinct arch, the preference set more than 400 years before Vitruvius even saw the light of day. The other signs of a Great Man are, "Long, slender fingers, Pliant hands and feet, full sized heels, Thighs like a royal stag, Full round shoulders, Hands reaching below the knees, Well-retracted male organ, Height and stretch of arms equal, Every hair-root dark colored, Body hair graceful and curly, Golden-hued body, Ten-foot aura around him. Soft, smooth skin, Soles, palms, shoulders, and crown of head well-rounded, area below armpits well-filled, Lion-shaped body. Body erect and upright, Full, round shoulders, Forty teeth, Teeth white, even, and close, Four canine teeth pure white, Jaw like a lion, Saliva that improves the taste of all food, Tongue long and broad, Voice deep and resonant, Eyes deep blue, Eyelashes like a royal bull, White curl [a spiral or circular dot] that emits light between eyebrows, Fleshy protuberance on the crown of the head."

For centuries after his death, the Buddha's followers considered it sacrilegious to produce full images of him. One acceptable alternative appears to have been a footprint, actually, a model showing the bottom of the foot. The surface was often decorated with Buddhist symbols such as the 1,000-spoke wheel representing the thousand teachings of the Buddha and the swastika (from the Sanskrit words *su* meaning *well* and *asti* meaning *to be*), a universal symbol of life and health even older than the Egyptian symbol of life, the ankh. Early on, the clockwise (arms facing right) swastika and its opposite, the counter-clockwise sauvistika, were interchangeable, but then the Nazis appropriated the clockwise swastika as their own, turning it into a symbol of death.

Christians embraced the arched foot in 325 when the Emperor Constantine (c. 272–337) convened the Council of Nicea one year after declaring Christianity to be the official religion of the Roman Empire. As conquerors often do, Constantine immediately set about sacking the losers' Greek and Roman temples and attempting to discredit the ancient pagan gods.

One particular target was Pan. The Greek word *pan* means *all*, and although Zeus was the King of the Olympian gods, Pan—often called The Great God Pan—was the protector of all things natural and wild. Half man, half goat, with hoofs and horns, Pan was musical and sexual, so much so that his name was used to characterize states of emotional excess: panic and pandemonium (the god's name plus the Greek word *daimon* meaning *evil spirit*). Naturally, this made him extremely popular, reason enough for the Christians to turn him into the model for their Devil, his cloven hoofs the Devil's feet. One unintended consequence of this fascination with the Devil's foot was a skeptical view of shoes and shoemakers. In Plato's *Republic,* Socrates names shoemakers as one group of tradesmen required to satisfy

PAN.

the "bodily wants" of citizens of the ideal state, but in Medieval Europe, people considered foot coverings (and the people who made them) suspicious because they could hide a cloven hoof. Or a flatfoot, considered an equally suspicious mark of a disciple of Pan. Eventually, the boisterous and joyful Greek deity was transformed into the patron of witches and visible flatfeet into a clear mark of allegiance to Pan that came in handy when nominating candidates for burning.

Eventually, people pushed the significance of various body parts beyond folklore into the realm of pseudo-science, proposing entire classification systems that would make it easy to identify superior and inferior races, as well as exemplary or dangerous individuals, by their physical characteristics, head to toe.

For starters, there was phrenology, from the Greek words *phrenos* meaning *mind* and *logos* meaning *study*. This scheme was the brainchild—yes, pun intended—of a German physician named Franz Joseph Gall (1758–1828). Noting that the human cerebral cortex, the wrinkled "gray matter" covering the brain, was larger than that of animals, Gall correctly identified it as the site of intelligence and personality. But then he went one step further off the cerebral cliff, insisting that the shape of the skull mirrored the shape of the cortex whose bumps and grooves and folds (*sulci*) he considered indicative of specific character traits, good, bad, and morally neutral. Gall was particularly taken with a number of bumps behind the ears, which he associated with nasty behavior such as thievery and deception; he associated other irregularities on the surface of the cranium with everything from loving one's children to an intention to murder. After his death, Gall's followers kept his ideas alive into the early twentieth century, but the theory faded from view as advances in surgery, neurology, and medical technology such as the encephalogram that records brainwaves, magnetic resonance imaging (MRI), and positron emission tomography (PET) permitted researchers to identify the actual loci of brain function and watch the path of neuronal activity in the living brain.

For Gall, the skull was all. Alphonse Bertillon (1853–1914) took a larger view, using the physical structure of the entire body to predict behavior, criminal and otherwise. Bertillon was not a physician. He was a clerk employed by the Paris police, but he was a clerk with vision sharp enough to propose that anthropometry, the careful measurement of body parts, might be one way conclusively to tell one person from another. Modern anthropologists use anthropometry to describe and date traces of living things such as their bones or footprints. Bertillon wanted to use it to identify criminals. In 1888, the Parisians were so enchanted with the possibility that they made him head of a newly created department of judicial identity and gave him full permission to test his theories. Things were going swimmingly until the introduction of fingerprinting showed that two people could have exactly the same body measurements, but different fingerprints, thus eliminating anthropometry as a law enforcement tool.

But the reliance on physical characterizations to label or libel people continued apace.

As the Europeans moved into Africa, one obvious difference between the colonized and the colonizers was skin color, while back home the shape and size of the nose remained an accepted sign of Other-ness. So did a narrow chest, now considered a predictor of weakness leading to tuberculosis. And the flatfoot was always fair game, even in fairytales.

The Brothers Grimm, Jacob (1785–1863) and Wilhelm (1786–1859), were German academics and folklorists. Their first book, *Children's and Household Tales* (*Kinder- und Hausmärchen*) appeared in 1812. Tucked in among the stories was one called *The Three Spinners,* in which the sighting of a flatfoot spared a princess from having to spin flax into thread, a chore she loathed. It goes like this: A young girl who refused to do household chores was sent by her mother to live with a queen who showed her to a room filled with flax to be spun into thread. The queen promised her oldest son as a prize when the girl finished spinning the flax. Lacking the skill and the will to do the job, the girl goes to the window to look out, maybe even to climb out. Spying three odd-looking spinner women on the street below, she invites them up. They are not a pretty trio. The first has a large lower lip from moistening the flax; the second, a wide thumb from twisting the thread; and the third, a broad flatfoot from working the peddle on the spinning wheel. Bamboozled by the girl's promise of an invitation to her royal wedding if they agree to spin the flax, the ladies go to work. They do the job. She reneges on her promise, and her prince, repelled by the spinners' deformities, declares that his new princess shall never put hand to spinning wheel again.

It's a happy ending for her, but not the end of the flatfoot as bad news.

Beginning in the 1930s, more than twenty-three centuries after Lycurgus, Plato, and the Decemvirs endorsed the elimination of infants with birth defects as a way to preserve the sanctity of "the guardians breed," the National Socialist German Workers' Party (*Nationalsozialistische Deutsche Arbeiterpartei*)—Nazis for short—adopted a program of murder in the name of racial purity, targeting adults who qualified as Other, a category whose descriptors included physique, physiognomy, birthplace, and, of course, religion.

By the last decades of the nineteenth century, the flatfoot was the "Jewish foot," a linkage that survived in early twentieth century medicine and as a tenet of Nazi propaganda. In *Der Giftpilz* (*The Poisoned Mushroom*), a children's book written by Julius Streicher (1885–1946), founder and publisher of the notorious anti-Semitic weekly tabloid *Der Sturmer* (*The Attacker*), one student, described as the best in the class, earns his teacher's praise by explaining that "Jews are usually: Middle sized and have short legs. . . . Jews have curved legs and are flatfooted."

After World War II, Streicher was tried for "crimes against humanity" by the International Military Tribunal at the Nuremberg War Crimes Trial. The indictment against him read in part, "In his speeches and articles, week after week, month after month, he infected the German mind with the virus of anti-Semitism, and incited the German people to active persecution." Streicher was convicted on October 1, 1946, and hanged two weeks later.

Jews weren't the only ones tagged with flatfeet. Across the Atlantic, smack in the middle of a map of Hanover County in northeastern Virginia was a place called Negro Foot (sometimes Negrofoot), one of literally dozens, maybe hundreds, of spots throughout the American South whose names once included the offensive N-word. This particular enclave was home to Patrick Henry when he issued his famous challenge—"Give me liberty or give me death"—at St. John's Episcopal Church in Richmond and while he was elected governor of Virginia in 1776. Ms. Dolley Payne, better known as Mrs. James Madison, also called it home. Today, the town is said to be virtually uninhabited.

In *Podunk: Ramblin' to America's Small Places in a Dilapidated Delta 88* (2012), a journal of his travels to America's "out-of-the-way communities," author Peter Zimmerman offers three accounts of how Negro Foot was named. The first two deal with gruesomely unpleasant racial encounters leading to amputated feet. The third is equally biased, but barely more palatable. In 1716, just about one-hundred years after the first African slaves landed in Virginia in 1619, the British royal governor Alexander Spotswood (1676–1740) was traveling with his retinue of local nobs, Native Americans, soldiers, and the requisite servants across the Blue Ridge Mountains to explore the Shenandoah Valley. As the company marched along they suddenly saw in their path the print of a barefoot "which evidenced the flat arch and the broad toe span that belong only to the negroid races" and straightaway named the place where the footprint appeared. In keeping with the prejudices of the time, he gave it the unpleasant N-word name that appears to have remained on U.S. Department of Interior maps as late as

1989. But the times, as Bob Dylan told us, certainly are a-changin'. In October 2011, an OpEd in the *Richmond Times Dispatch* by editor/columnist A. Barton Hinkle cited a number of such places—"a couple of Negro Hollows, three Negro Points, four Negro Runs, the Negrohead summit in Rockingham"—and the modern move to re-name them, under a headline asking the salient question, "Would you ask a friend to lunch at Negro Foot?"

Spotswood was hardly the only one to classify flatfeet as an African American trait. Even so vocal an opponent of slavery as Julia Ward Howe (1819–1910), author of *The Battle Hymn of the Republic*, wrote in *A Trip to Cuba* (1859) that "the negro among negroes is a coarse, grinning, flat-footed, thick-skulled creature." The sentiment, widely shared, gave opponents the opportunity to attack both Howe and her fellow abolitionists as hypocrites.

So imagine everyone's surprise when, late in the twentieth century, three separate studies over a period of seven years appeared to confirm a link between ethnicity and flatfeet.

In 2003, a team of public health epidemiologists from New England Research Institutes, Boston University School of Medicine, New England Baptist Hospital, and Sargent College funded by the National Institutes of Health's Institute on Aging set up a "community-based, multiethnic (non-Hispanic White, African-American, and Puerto Rican) random sample of 784 community-dwelling adults aged sixty five or more years in 2001–2002 in Springfield, Massachusetts" to determine "The Prevalence of Foot and Ankle Conditions in a Multiethnic Community Sample of Older adults." Telephone and in-home interviews plus orthopedic examinations showed an overall incidence of flatfeet (19 percent) matching most common estimates of 20 percent of the population, with similar figures for men and women and people of varying education levels. The one significant marker was race: Flatfeet were most common among African Americans, followed by non-Hispanic Whites and Puerto Ricans. High arches, found in 5.2 percent of the volunteers, were more common among women than among men, but the percentages did not differ by race.

Six years later, a report in *Foot and Ankle International* by a team of Costa Rican investigators showed a similar racial variation in two angles of the bones in the foot, each variation associated with either a higher or lower risk of flatfoot. The first angle, calcaneal pitch (CP), is visualized by drawing a horizontal line along the bottom of the foot and another line tracing the rise of the foot bones from the bottom of the heel: the lower the pitch, the flatter the foot; the higher the angle, the higher the arch. The second angle, the lateral talocalcaneal angle (TCA), is shown with a similar diagram, this time between the bottom of the heel and the ankle bone: here too, the lower the angle, the flatter the foot. Measuring the feet of 126 volunteers—forty-five African Americans, fifty-nine Caucasians, twenty-two Hispanics—the study found "significant" differences by race. As a rule, the African Americans had lower CP angles than either the Caucasians or the Hispanics; conversely, the Caucasians had higher TCA angles than the African Americans.

Finally, in 2010, arthritis researchers from the University of North Carolina at Chapel Hill released data from a four-year survey of 1,536 volunteers designed to compare the frequency of foot disorders among African Americans versus Caucasians. Their study, "Racial Differences in Foot Disorders: The Johnston County Osteoarthritis Project," showed that African Americans older than forty-five were three times more likely than Caucasians of the same age to have flatfeet, while the Caucasians appeared five times as likely to have high arches and insteps. Bunions and hammertoes were equal opportunity problems. "That suggests there is a real racial difference," said the study's lead author, Yvonne M. Golightly. "The next step in our research is to determine the origin of these disorders."

Perhaps owing to the absence of Asians from Boston, Costa Rica, and North Carolina, these studies have no statistics on the incidence of Asian arches, high or low, but at least one shoemaker was literally banking on the former. In 2011, British designer Rupert Sanderson created a shoe with a four-inch high heel and extra padding at the instep to fit the presumably higher Asian female arch. Unfortunately, his design, based on a mold of the highly arched foot of one female friend, contradicts podiatric studies which suggest that Asian feet are slightly broader in front and lower in the arch than Western European feet and that the incidence of flatfeet among Asians may run as high as 80 percent, the exact opposite of Western feet. But the good news is that high arch or low, the extra padding raises the foot and makes the high heel feel lower and less stressful.

Eventually, Western sensitivity to ethnic and religious slurs changed "Jewish foot" to "weak foot," but a flat arch was still considered reason enough to exclude men from the armed forces, the ultimate protector from the Other. Around the world, the military exhibited a serious interest in the state of a recruit's or conscript's feet. When you think of it, this is probably not surprising considering that in pre-mechanized times, an Army really did march on its feet.

On November 15,1863, *The New York Times* published a list of 41 "Disease and Infirmities" the War Department considered serious enough to guarantee exemption from the draft for the Union Army. In addition to the usual imbecility, insanity, epilepsy, paralysis, habitual and confirmed intemperance or "solitary vice," incontinence, loss of a limb or a finger, deafness, blindness, a curved spine, excessive obesity and so on, there was this: "Club feet; total loss of a great toe. Other permanent defects or deformities of the feet, such as will necessarily prevent marching." That may or may not have meant flatfeet. Either way, it was number 39 on the list, clearly not a high priority item.

The British, on the other hand, were positively obsessed with problem feet. In 1989, two researchers at the British Military Hospital at Rinteln, West Germany, tracked the British military's treatment of flat-footed servicemen over a period of 300 years. In their paper "The Longstanding Problem of Flatfeet" (note the subtle play on words; people with flatfeet may indeed have trouble standing or marching comfortably for long periods), they include an Army order in 1690 directing that "[f]or the exercise of the musket . . . the musket being shoulder'd, the feet are to be at one step distance, the heels in a straight line, the toes a little turn'd outwards."

Fast forward 200 years, and the flatfoot problem becomes one of dollars and cents, or, in this case, pounds and pence. After World War I, British bean counters auditing medical discharges found that "[i]n 1922–23, each recruit enlisted and subsequently discharged after a period of training cost the State 50 pounds, and discharges for flatfeet accounted for 8.5% of all medical discharges during training."

For years, the armies of the world waged their war on flatfeet, turning down men and women whose lack of arches actually posed no problem in favor of those with discernible arches whose common but hidden disabilities often surfaced during basic training. In 1901, when Albert Einstein (1879–1955) was called up for military service in the Swiss army, his feet ruled him out; fifty-nine years later, author Stephen King's Maine draft board turned him down for (among other things) the same reason. Barred from the army, Einstein became a technical expert at the Bern Patent Office at 3,500 francs per year; in his spare time, he pursued his studies in theoretical physics, on his way to the Nobel Prize in 1922. As for King, a supporter of the anti-Vietnam War movement, he went home to become a high school English teacher and write the stories and books that earned him several millions more than Albert Einstein's 3,500 francs a year plus the National Book Foundation's Medal for Distinguished Contribution to American Letters (2003). As he has often said, King owed his success at least in part to his wife Tabitha, who in a moment of true literary prescience, is said to have fished the first pages of the manuscript for *Carrie* out of the garbage where King had tossed it and then convinced him to finish what went on to be his first published novel, selling more than one million copies in the first year after publication in 1974.

Physicists and novelists weren't the only ones rejected, or depending on your point of view, saved by their arches.

During the Vietnam era, future Arkansas governor Mike Huckabee enrolled in the Reserve Officers Training Corps (ROTC) in college and expected to serve in combat, but was ruled out by his feet, as was South Dakotan Tom Brokaw, who wanted to serve in the Navy, and applied to Officer Candidate School. Boston-born New York mayor Michael Bloomberg was rejected by the Army. Flatfooted Newt Gingrich, a hawk on Vietnam, was a nearsighted graduate student with several children who said his deferments didn't matter because he "didn't think one person would make a difference." But the man with the best flatfoot deferment story is Charles Dubin, a television director who was rejected by the Army during WWII and then blacklisted after the war in 1958 for refusing to give names to the House Un-American Activities Committee. After a few bumpy years, Dubin went on to become the person who directed the largest number of episodes of the hugely popular, hugely anti-war television show *M\*A\*S\*H*, including one in 1979 titled "Are You Now, Margaret?" in which Hot Lips Houlihan (Loretta Swit) was pressured by a Congressional aide to name friends she knew to be Communists. Margaret just said, "No," thus proving that for Dubin revenge against a government that had twice dissed him—first for his flatfeet and then for his presumed disloyalty—was twenty-one years later indeed "a dish best eaten cold." Today, the U.S. Department of Defense only turns down men and women with "pronounced cases [of] . . . decided eversion [turning outward] of the foot and marked bulging of the inner border, due to rotation of the talus [ankle bone], regardless of the presence or absence of symptoms."

Like various armed forces, American cops have had a long-standing semantic relationship with flatfeet. My *Merriam-Webster New Collegiate Dictionary* dates the first use of the term "flatty" as shorthand for a police officer to 1899 in Akron, Ohio, where it was used to distinguish foot patrolmen from officers riding in the recently introduced electric motor-powered patrol wagons. Actually, *Merriam* missed the mark; *The National Police Gazette* had used the term thirty-three years earlier, reporting that some miscreants were "pulled in by a flatty cop." In 1911, Damon

Runyon (1880–1946) published his first book, a collection of poems titled *The Tents of Trouble (Ballads of the Wanderbund, and Other Verse)*, which included this line: "We croaked a flatty in Baltimore and we beat, by a nose, the law." And in 1913, Canadian novelist, screenwriter and poet Arthur Stringer (1874-1950) wrote a mystery titled, *The Shadow* (no relation to Lamont Cranston, hero of the later radio program that ended each week warning that "The weed of crime bears bitter fruit. Crime does not pay. . . . The Shadow knows!") featuring "Never-Fail" Blake, a detective hero who dealt frequently with "confident striding 'flatties' with their ash night-sticks."

Foot patrolmen weren't the only law enforcement officers identified by their feet or foot coverings. The word *gumshoe* originally meant simply a shoe with a rubber sole, but by the turn of the twentieth century, it had evolved into a nickname for a detective—also known as a dick, an operative, a private eye, a P.I. (private investigator), a Sherlock, or a shamus, this last perhaps a corruption of the name Seamus and an obvious nickname for a profession then dominated by the Irish—because the shoes allowed a man to walk quietly, following his prey without being noticed. For the same reason, the canvas-topped rubber-soled shoes introduced in 1916 by the U.S. Rubber Company under the Keds brand name were christened *sneakers* by Henry Nelson McKinney, an executive with N.W. Ayer & Son, the agency that created the ads for the shoes.

There is even a kids' version of a crime-fighting flatfoot: *Flatfoot Fox*, "the smartest detective in the whole world." The canny Fox, hero of five books by Clifford Eth, the pseudonym of Eth Clifford Rosenberg, a well-regarded children's author and the biographer of Yiddish actress Molly Picon. Today, fifty-seven boxes of her papers, including the copy-edited proofs of some of the Flatfoot Fox books, are stored at the University of Minnesota Libraries Children's Literature Research Collections. Rosenberg's Fox solved cases in the animal world with the help of a friendly Secretary Bird, which in real life is a fairly scary animal, a raptor with long legs, wings, and tail, named for a crest of long feathers atop its head that look like old-fashioned quill pens.

Back on the mean streets of the Big City, one early twentieth century member of a profession definitely not meant for children was also nicknamed "flatfoot," courtesy of jazz instrumentalist and composer Bulee "Slim" Gaillard (1916–1991). Gaillard, who spoke eight languages and invented one of his own, a dialect he called *Vout*, published "Flatfoot Floogie with a Floy Floy" in 1938. The song was originally titled "Flatfoot

Floozie"—from "Flossie," early slang for a promiscuous girl–but the name was changed to avoid the radio censors' blue pencil. The term *Floy Floy* slid by under their radar, un-noted except by those who knew quite well that it was slang for a venereal disease, probably syphilis whose third-stage symptoms include tabes dorsalis, the degeneration of nerve tissue that leads to loss of muscle and tendon reflexes in affected limbs, causing a lurching gait that seems to be referenced in a line from the last verse of the song: "If you go on stumblin' with the gang!"

Coded language and all, "Flatfoot Floogie" was recorded by such household names as Louis Armstrong, The Mills Brothers, Benny Goodman, Django Reinhardt, Count Basie, Fats Waller, and the luminous Nat King Cole. Not one of them publicly batted an eye as hep cats of all ages jitterbugged to a tune about a prostitute with late-stage syphilis, just as thirty years later their hippie children and grandchildren embraced "Puff, The Magic Dragon" and "Lucy in the Sky with Diamonds" that everyone—except the BBC, which banned "Lucy"—agreed were definitely, absolutely not code for marijuana or lysergic acid/LSD. The hep cats and the hippies also had this in common: According to the *Online Etymology Dictionary*, both sobriquets come from *hep*, first used as a synonym for "up to date" in a 1908 edition of *The Saturday Evening Post*. *Hep* itself is variously described as being either underworld slang, or as some etymologists suggest, a semantic descendent of *hipi*, a word in Wolof, the language spoken by as many as seven million people in West Africa, that means *to see* or *to open your eyes* or *a person with open eyes.*

"Flatfoot Floogie" was Number One on the *Hit Parade* for eight weeks. Another Gaillard song, "Cement Mixer, Putti, Putti" (1946) made the *Hit Parade*, but "Flatfoot Floogie" was the true winner, so popular that a copy was included in a time capsule buried at the site of the 1939 New York World's Fair to be opened in 5,000 years.

Gaillard told friends he intended to be around for the occasion.

## Fixing or not fixing the no-problem problem

Despite the social, religious, and military obsession with flatfeet, the good news is that people born with the flexible version rarely experience serious discomfort. The so-so news is that there has been very little serious investigation into what causes flatfeet, who is most likely to have them, and what if anything to do about the supposed problem.

Like clubfoot and gout, the tendency to flexible flatfeet and high arches sometimes runs in families. Unlike clubfoot, however, there is no known link to a specific gene, and unlike gout, flatfeet are not linked to any metabolic disorder. In fact, when pressed, some experts suggest that any problem with flattened arches may be mostly in the mind of the observer not in the flatfooted person's feet.

In 1989, when the Army Institute of Environmental Medicine (Natick, Mass.) ran a study of more than 300 Army infantry trainees at Fort Benning, Georgia, they came to three fascinating and totally non-alarming conclusions.

Estimates of the number of flatfooted soldiers in the cohort ranged from a low of 10 percent to a high of more than 50 percent, depending on which of the six doctors running the study did the evaluating. To no one's surprise, different researchers had different ideas about exactly what a flatfoot looked like, so the authors took pictures of every single one of the 600 feet in the study and did dozens of complicated measurements to visualize the actual structure of the arch in each foot. Then, because most estimates put the percentage of flat footed people in the general population at two out of every 10 people, they simply called the 20 percent of the feet with the lowest arches "flat."

Having done this, they discovered that soldiers with feet that were obviously flat were *less* not *more* likely than those with normal feet to suffer an injury during basic training. And in a perfectly counterintuitive finding, it appeared that those with higher-than-normal arches and insteps were twice as likely to end up with sprains and stress fractures, possibly because a foot with a high arch is a less efficient shock-absorber.

Having said that, it is my duty to report that in 1999, a longitudinal (long-term, with follow-up) survey of 449 Navy SEALS run at the Naval Health Research Center in San Diego, California, and the Naval Hospital at Camp Lejeune in North Carolina produced similar results: Over time, those with flatfeet were at higher risk of stress fracture and inflammation of the Achilles tendon, the tissue that ties the calf muscle to the back of the heel. Nonetheless, the author of the Fort Benning study told *The New York Times* in 1990 that "[m]uch of what we've believed about flatfeet is mythology. I've seen drill sergeants with arches as convex as the bottoms of rocking chairs, who are active and successful." The same goes for professional athletes. "If I had a choice as a professional athlete to have a high arch or a flat foot, I'd take the flat foot," said Michael Coughlin, then-president of the American Orthopedic Foot and Ankle Society..

He was right. Since then, a large number of continuing studies still attest to flatfeet being a no-problem problem. For example, in 2006, Saudi

Arabian investigators at the Al-Hada and Taif Armed Force Hospitals in Taif, about 500 miles from the capital city of Ryadh and more or less 6756.5 miles (10873.5 kilometers) as the Boeing 777-300 flies from Washington D.C., ran a similar study and produced similar results among 2,100 recruits aged eighteen to twenty one, and got similarly much-less-than-alarming results. Once again, there was some disagreement as to exactly what constituted a flatfoot but those men whose feet everyone agreed were definitely flat often had relatives with the same condition, and flexible flatfeet were "of little consequence as a cause of disability in adults." The only surprise was that just 5 percent of the men in the study met the Saudi researchers' standard for having flatfeet versus the usual 20 percent cited in the United States and around the rest of the world. Perhaps it was the result of there being more than the usual number of differing ideas of what a flatfoot looked like.

As for children, in 2005, researchers at the University of New South Wales (Australia) tested the movements of fifty-four children, age nine to twelve, half with flatfeet. Each of the seventeen children with flatfeet was fitted with a device between foot and knee to monitor the movement of the child's joints to be measured and interpreted by computer. In the end, the flat-footed kids had balance and motor skills equal to children with normal arches, but were able to jump 15 percent higher, although how they did it with all those devices strapped on is certainly a mystery. The only area in which the flatfooted lagged was what Dara Twomey calls "lateral hopping ability," that is, hopping sideways over a piece of string. Like the children in the 2011 Aberdeen study of clubfoot, Twomey's young subjects took their assumed disability in stride. "I play netball and quite a bit of sport and I don't feel there's a difference or anything," one of them said. "Most of my friends haven't got flatfeet and I'm just as good as them in practically everything." And probably in shoes without the inserts often prescribed for people with flatfeet, because they may not be much of a treatment after all.

As always, the Greeks had a word for it: *orthos* meaning *straight*. But *orthotic*—a device to support a body part such as the arch in your foot—did not enter the English vocabulary until the 1950s (*Concise Oxford English Dictionary*), 1955 (*Merriam-Webster*), or 1960–1965 (*Unabridged/Random House Dictionary*), take your pick.

The first effort to cushion a foot inside a shoe was most likely layers of wool or another soft material piled inside a Greek or Roman sandal to relieve an aching less-than-perfect Greek or Roman arch. More than two thousand years later, in 1865, Everett Dunbar of Bridgewater, one of

Massachusetts' many shoe manufacturers, took a giant step forward by sticking leather lifts between inner and outer sole of his shoes to support the arch.

The first stand-alone arch support, the Foot-Eazer™, was designed by William Mathias Scholl (1882–1968) in 1903, one year before he graduated from Chicago Medical School, now the Rosalind Franklin University of Medicine and Science named in honor of the British biophysicist whose radiographic images of DNA were indispensable to James D. Watson and Francis Crick's discovery of the double helix. The following year, Scholl, the son of a German immigrant cobbler, was awarded U.S. Pat. No. 1575645 for his two-piece insert, leather on top of a rigid "silveroid" (a copper/nickel alloy also known as silverine, Alaska silver, or German silver) base, with a bump in the middle to support the arch. Scholl sold his insert himself, walking around to Chicago shoe stores toting along a foot from a human skeleton on which to demonstrate the insert, some of which survive to this day.[1]

Just as Scholl was beginning to peddle his insert in Chicago, Royal Whitman (1857–1946), an orthopedist at New York's Hospital for the Ruptured and Crippled, now the Hospital for Special Surgery, created his own arch supports, rigid, heavy metal devices called the Whitman Plates that were difficult to wear, but nonetheless widely prescribed by orthopedic surgeons. Whitman authored a now-classic paper, "Study of the Weak Foot, With Reference to Its Causes, Its Diagnosis, and Its Cure; with an Analysis of a Thousand Cases of So-Called Flat-Foot," in the *Journal of Bone & Joint Surgery* in 1896. In his article, Whitman laid out the differences between congenital flatfoot and flatfoot acquired as an adult. The first, he wrote, was "hereditary and in some areas a race-linked condition." The second was not true flatfoot, but "a minor element of weakness and a secondary element of the deformity; for the symptoms of flat-foot do not result because the foot is flat, but because it is becoming flat; they are the symptoms of strain up on the weak foot and of the injuries and changes accompanying a progressive dislocation." Whitman took exception to the usually casual approach to flatfeet. Like doctors treating clubfoot, he proposed bracing, casting, and sometimes surgery for flatfeet and weak feet—plus his own arch support.

Some of this was probably unnecessary.

Flatfeet may be uncomfortable, but they are rarely crippling. In fact, so long as they are not yours, they can seem comical, like the big flat shoes worn by circus clowns. That may be one reason why the index for *Harrison's Principles of Internal Medicine*, the text the *Journal of the American Medical Association* described in 2012 as "arguably the most recognized book in all of medicine," includes flat warts and flat worms, but not flatfeet.[2]

After the introduction of anesthesia and antiseptics made surgery safer, flatfeet, like clubfeet before them, became an inviting target for the newly empowered surgeons whose hands practically itched to realign ligaments, tendons, and bones in the attempt to improve a sagging arch. This kind of surgery may well benefit people with severely deformed flatfeet or feet made rigidly flat by abnormally fused bones, or flexible flatfeet that stiffen with age or adults with an acquired flatfoot (or flatfeet) due to a supporting tendon's tearing or slipping and falling and taking the arch along with it, but asymptomatic flatfeet, even in adulthood, seem best left uncut.

The other surgery for flexible flatfoot is sinus tarsi implant, a.k.a. subtalar arthroereisis, a procedure in which a small titanium device is inserted into a small hollow (sinus) in the foot between the heel bone (calcaneum) and the ankle bone (talus). It has not been greeted with enthusiastic applause. In 2009, the British National Institute for Health and Clinical

Excellence (NICE), an independent, government-funded organization that advises the British National Health Service (NHS), assessed the effectiveness of arch implants. They found a high rate of postsurgical complications such as pain, a shift in the position of the implant that required it to be removed, or on rare occasion, the actual exit of the implant from the foot.

Two years later, just as *The Merck Manual*, the continuing series of volumes offering orthodox American medicine diagnosis and treatment, chose to recommend surgery for some flatfeet, an article in *AAOS Now*, the journal of the American Academy of Orthopedic Surgeons, reported that although surgery might make sense for some cases of rigid flatfoot, flexible flatfoot is normal in babies and in about one out of four adults. Slipping an implant into one of these feet, the author decided, was a complication, not a solution; major insurance companies will not pay for the surgery as a treatment for either congenital or adult-acquired flatfoot because they contend it is an "experimental and investigational [procedure whose] . . . clinical value has not been established."

So thanks to Scholl and Whitman—but mostly Scholl, who eventually built a multi-million dollar business on top of his Foot-Eazer™—orthotics remain the treatment of choice. A worldwide industry now thrives based on Dr. Scholl's store-to-store sales strategy. Lighter materials, better fit, more comfortable support, claims of fewer injuries among athletes, support sales of inexpensive over-the-counter inserts to special customized ones costing several hundred dollars a pair.

The inserts are supposed to compensate for an irregularity in the structure of the foot, most commonly, a low arch, but neither orthopedists, nor podiatrists, nor neurologists, nor fitness experts, nor scientists specializing in biomechanics (the force exerted by muscles and gravity on our bones and joints) can tell you exactly how orthotics inside your shoes influence your movements or affect the stress you put on joints and muscles when you walk, run, skip, jump, hop, or do any one of the things you do while standing up.

Orthotics enthusiasts will tell you most studies suggest that people do feel better with inserts in their shoes, although which inserts produce what effect is definitely open to dispute. When Benno M. Nigg, professor of biomechanics and co-director of the Human Performance Lab at the University of Calgary in Alberta, asked one distance runner to test orthotics from five reputable manufacturers, each insert designed to correct his pronation in a different way, the runner voted for the two that seemed to enable him to run faster. But the two he chose had totally different structures.

Which led Nigg to ask, *do* orthotics work? And if so, how? That pretty much depends on who's wearing the insert. One person using the arch may end up walking on the outer edge of her foot and maybe even grabbing at the floor with her toes in an effort to hold her balance, a second might just continue along in the position his muscles and joints consider normal even with the insert in place, and a third just might be positioning his foot more normally with the arch than without it.

What's a flatfooted person to believe?

The same Old Wives who prescribe chicken soup for a cold say that flat-feet are a sign of a bad temper. They also say that it's bad luck to allow your "first footer" (an old-fashioned term for the first person to walk through your door after the stroke of midnight on New Year's Eve) to be someone who has eyebrows that meet in the middle or cross-eyes, or, what else? a person with flatfeet.

The chicken soup gained a certain legitimacy in 2000 when a team of pulmonary specialists at the University of Nebraska Medical Center found a wealth of honest-to-goodness anti-inflammatory compounds in chick-en, onion, sweet potato, turnip, carrot, celery, and parsley in broth with no salt, but lots of matzoh balls. Their report, in the journal *Chest*, said that the soup helped reduce the movement of neutrophils, anti-viral and anti-bacterial cells that multiply during an infection to behave like tiny vacuum cleaners, sucking up cellular debris. To validate their findings, the lung doctors also tested commercial chicken soups (some were more active than the homemade version), vegetable soup (minor effect), and plain tap water (no effect). Clearly, the star of the show was the chicken soup, with or without the matzoh balls.

There's no such proof of a connection between flatfeet and bad luck or a bad temper, but it would not be unreasonable to assume that a person with a physical characteristic other people consider other-ly might be a bit cranky from time to time. Or at least until he or she learns that flatfeet may well have played an important role in setting the rules by which we measure our world.

## Foot rules and rulers

Sometimes our words have more than one meaning. For example, when the Greek sophist Protagoras (c. 490–420 BCE) said that "man is the mea-sure of all things," he meant that we, not our gods, determine our fate. But there is another way to interpret what he said, because man really is the measure of his world.

Measurements, Secretary of State John Quincy Adams wrote in a report to Congress in 1817, "are among the necessities of life to every individual in society." Length—the height of a man or the distance from here to there—is particularly important. From the beginning, every ancient society had a way to measure it and virtually all the values began with our anatomy: our fingers, our hands, and our feet. This link is so strong that in many Romance languages the word for inch and thumb are the same or come from the same root. In French, the word for both is *pouce*; in Italian, *pollice*; in Dutch, *duim*. In Spanish, an inch is *pulgada*, a thumb, *pulgar*; in Portuguese, it's *polgada* and *polegar*.

In China, the legendary Huangdi (Yellow Emperor), who ruled in the area of the Yellow River sometime between 2697 and 2590 BCE, is credited with creating language characters, ships, carts, medicine, and music and a system of measures whose smallest standard unit was the width of a person's thumb at the knuckle. To eliminate variation, and the inevitable question—"Whose finger?"—the system was standardized relatively quickly under another legendary leader, Yu The Great, (c. 2200–2100 BCE), whose other claim to fame was to prevent flooding of the Yellow River by dredging channels to send the overflow waters east into the Bohai Sea and from there into the Yellow Sea.

Among citizens along the Mediterranean and in Mesopotamia, the area of southwest Asia that corresponds roughly to modern Iraq, the basic unit of length was the *cubit* (from the Latin *cubitum* meaning *elbow*), roughly the length of an adult (usually male) forearm from elbow to outstretched fingertip, a distance between 18–22 inches, depending on whose forearm one used. Arabs, Egyptians, Greeks, Jews, and Persians divided their cubits into smaller units, also based on body parts and also highly variable. For all of them, the smallest unit was the digit, always roughly an inch long, but never exactly so. The Arab digit appears to have been 0.9 inches; the Hebrew, 0.75 inches; the Persian, 0.8. The next unit up, the palm, was measured in multiples of finger widths, and after that, the cubit in multiples of palms. In Rome, for example, four *digiti* (finger widths) equaled one *palmus*; four *palmi*–"i" is the plural of the Latin masculine singular "us"—or sixteen *digiti* equaled one *pes* (foot).

Other body-based measurements of length include the yard (3 feet), once put at the distance from the tip of your nose to the end of the middle finger on your outstretched arm; the span (nine inches), the width of a fully extended hand measured from the tip of the thumb to the tip of the little finger; the fathom, the distance from the tip of the middle finger on one outstretched arm straight across your body to the tip of the middle finger on the other outstretched arm; and the mile, from the Latin *mille passuum* or 1,000 paces, a pace in Rome being equal to five pes, making the original Roman mile 4,867 feet long ($11.68 \times 5 \times 1,000 / 12$).

The differences from system to system, so small at the digit stage, widened as the length increased, often enough to affect important details such as an architect's ability to assume the size of the materials required to build a building or an army's determination of how far in the distance the enemy waited. The Egyptians solved the problem with measuring sticks, wooden rods, or bars of exact length. The originals of these highly prized instruments were kept in the local equivalent of City Hall or City Temple and made available for copying, so that duplicates could be distributed to ensure that a cubit here was also a cubit there. Then the Romans further standardized the rules by introducing the *uncia,* which means one-twelfth and which we translate as *inch.* By common consent, the early Roman *digiti* is believed to have been about 0.73 inches long; the *palmus,* 2.9 inches; and the foot—the *pes*—about 11.68 inches, not quite 12, but close enough. After that, the concept of the twelve-inch *pes* gained increasing acceptance, and obviously still rules most of our measuring rules.

With some local tweaks.

The Britons, who used the furlong (660 feet) to measure property, changed the Roman mile of 1,000 *pes* to the British mile: eight furlongs or 5,280 feet, a move legalized by Parliament in 1592. Before the Norman Conquest, Anglo Saxons had their own variety of foot lengths including one equal to 12 inches, one equal to 13 inches (two shaftments), and the pes naturalis (natural foot), equal to slightly less than 10 inches, about the length of a foot that fits a modern U.S. size 9.5 or 10 shoe. The 12-inch foot, arriving with the Normans in 1066, is said to have been certified 50 years later during the reign of Henry I, along with the 3-foot long yard, described as "the measure of the king's own arm."

Clearly, each of these measurements, including the nearly 12-inch Roman *pes,* depended on the person serving as a model having been a fairly tall male. But when the Romans and their mathematically inclined descendents were measuring things in the length of some man's fingers and feet, exactly how tall was a "fairly tall" man?

Remember, based on the Vitruvian proportions, the ideal man's foot is exactly one sixth as long as he is high. Measure the foot in any drawing of Leonardo's *Vitruvian Man,* then measure the man's height, and you will find that, yes, the height of the figure is six times the length of the foot. But the Roman human foot was obviously shorter than the pes because an ideal *Vitruvian Man* with an 11.68-inch foot—exactly the length of the foot that fits a modern American size 13 shoe—would be only slightly less than 6 feet tall ($6 \times 11.68 = 70.08$ inches).

Today, the United States has no shortage of men whose height tops 6 feet and whose shoe size may run well into the teens. Or larger. At 7 feet 1 inches tall, weighing 325 well-buffed pounds, Shaquille O'Neal wears a size 22 shoe—15 and 11/16 inches long to fit over his 14 and 11/16 inch-long feet that are almost but not quite one-sixth of his height.

A slew of anthropometric studies calculating human body measurements in earlier times in order to identify and compare anthropological remains puts the normal height of persons living in Europe and around the Mediterranean neighborhood during the time of the ancient Romans at about 5 feet 3 inches. True, Julius Caesar and his grandnephew Caius Julius Caesar Octavianus, better known simply as Octavius, the first emperor of the Roman Empire, are both said to have stood a majestic 5 feet 7 inches. But even they would not have had that size 13 foot.

So here's the question: Was the Roman man whose foot became the measure of the pes not only a legend but also a giant in his own time?

Or were his feet simply pes planus, lacking a normal arch when standing and thus leaving a longer impression on the floor.

In short, did Leonardo's perfect *Vitruvian Man* have perfectly flat feet?

## Notes

(1)  In December 2011, Ebay item number 290466164613 offered a "Vintage Dr.Scholl's Foot-Eazer," marked "Dr. Scholl's Foot-Eazer. Pat. No. 1575645, Silveroid Made in U.S.A." for $9.99 plus $6 shipping. Minus the skeleton foot.

(2)  From the first edition in 1889 of *The Merck Manual*, straight through to the fifth edition in 1923, the series pretty much comprised catalogues of materia medica, that is, remedies, including Merck products. The *Merck* editors did not even mention flatfeet until the eighth edition (1950) when the recommended treatment for pes valgo planus (flatfoot) was "correction of the anatomic deformities through use of any of the following: adhesive strapping, arch supports, external shoe corrections (i.e. Denverheel and metatarsal bars), padding of insole with felt, operative intervention." For pes cavus (high-arched foot), it was "a sponge rubber bar placed across an insole just behind the metatarsal heads, or a leather bar applied similarly to the outside of the sole of the shoe." The ninth edition (1956) added warm soaks, hot and cold foot baths, and aspirin for flatfoot. The tenth edition (1961) translated aspirin into a more sophisticated "acetylsalicylic acid." To

deal with any discomfort, Merck still advised keeping the foot "in a relatively normal position while walking by an arch support made of sponge rubber," but added that "[m]etal or plastic is used for older people with concomitant tarsal osteoarthritis." In the fourteenth edition (1982), the words pes planovalgus are out. There is nary a mention of flatfeet, but cavus, highly arched feet, and pronation (inward leaning of the foot) are prominent in the entry on "metatarsal-phalangeal articulation pain (lesser toes)" including "excessive eversion of the subtalar joint (rolling in of the ankles)" for which "[o]rthotics should be prescribed to redistribute and relieve pressure from the affected articulation." Five years later, pes cavus and planovalgus are back in the index for the fifteenth edition (1987), to be treated with corrective shoes, including arch supports. By the eighteenth edition (2006), all these terms are gone, gone, gone, replaced by the *really* sophisticated *talipes equinovarus and talipes calcaneovalgus*; casting and surgery—the Ponseti duo—have been added to the list of possible remedies. Ditto for the nineteenth edition (2011).

# 4

# DIET

"A man hath no better thing under the sun, than to eat,
and to drink, and to be merry."

Song of Solomon

## HOW DO YOU SCULPT A HUMAN FOOT?

The question is more complicated than you might think. In 1897, E. H. Bradford, a Boston doctor, founding member and one-time president of the American Orthopedic Association, wrote that, "Anyone whose attention has been called to the subject of the shape of the human foot will find it of interest to examine the feet in a collection of statues." Having done so, Dr. Bradford concluded that every sculptor since the dawn of time had each chosen one of three types of feet.

The "barbaric and Egyptian type," he wrote, was a straight foot with straight toes and no natural curves, modeled on people who went barefoot or wore sandals. The "classical type" found in Greek, Roman, and modern

(i.e., late nineteenth century) statues was modeled on people who either went barefoot or "wore the cothurnus," a laced high boot with a thick sole, sometimes called a buskin and often worn by actors. This foot had the proper curves, with the big toe slightly separated from the other four. Finally, the modern pseudo-classical type, based on the feet of the "races that have worn shoes," was similar to the classical, but shorter and fatter with the big toe bent in toward the other four.

As you can see, in all three drawings the second toe is longer than the big toe.

Now look at the toes on the right foot of the *Vitruvian Man*. Unlike the toes on the ideal Buddha's foot, all exactly the same length, the toes in Leonardo's drawing are more true to life, or at least true to the life of the three of every four human beings whose genes dictate that the first toe should be larger and longer than the other four. One in five of us gets a different genetic message, making the second toe longer than the first, a feature clearly visible on the foot of God in Michelangelo's *Creation of Adam* panel on the ceiling of the Sistine Chapel (whether this represents the reality of Michelangelo's own foot or that of a male model is a mystery). Three in every one hundred humans have a longer third toe. Only two of every hundred have toes that look like those on the Buddha's foot with at least the first three equal in length.

These patterns of toe length occur on both male and female feet, but finger length is sexually dimorphic, meaning it is gender-related, different in men and women. At birth, a baby boy's ring finger is usually longer than his index finger; for baby girls, the reverse is true. In 1998, John Manning of the School of Biological Sciences at the University of Liverpool attributed this difference to the infant's exposure to sex hormones in the womb. In 2011, by manipulating the levels of testosterone and estrogen available to mouse fetuses, Zhengui Zheng and Martin Cohn of the Howard Hughes Medical Institute and the Department of Molecular Genetics and Micro-biology at the University of Florida College of Medicine were able to demonstrate that Manning was right. Prenatal exposure to testosterone appears to turn on genes that increase cell division in the fourth toe on the mouse's five toe-back paw, a digit comparable to our human ring finger, making it longer than the second toe, comparable to our index finger. Estrogen does the opposite, slowing cell division and producing a shorter fourth toe.

This was a fascinating discovery. Few take it seriously when palm readers and fortune tellers link finger length to specific characteristics such as aggression or infertility, but when reputable scientists suggest the same thing . . . that matters. After Manning's studies, several hundred serious medical researchers began to investigate the possibility that there might actually be a link between the index-to-ring-finger-size ratios and a plethora of human health conditions ranging from autism to sperm counts, sexual orientation, gender-related cancers, and, of course, athletic prowess. No such links have yet been proven, but Zheng and Cohn's data does open the door to a more reliable understanding of fetal exposure to chemicals in utero and thus to better predictions of fetal (and maybe adult) health.

Length aside, the big toe is always wider and sturdier than the others, and that is also a big deal.

Along with three firm bone-and-muscle arches, the big toe stabilizes your foot and literally puts the spring in your step. Begin to move ahead, and your foot flexes at joints that connect the long metatarsal bones in the middle of your foot to the shorter bones at the back of your toes. The entire foot bends forward, and your big toe, assisted by the four smaller ones, pushes you off the ground and sends you on, a subject of major importance to those who study biomechanics, the science of the ways in which your body moves. This discipline advanced by leaps and bounds at the beginning of the twentieth century due to the invention of motion pictures, which made it possible to film and then analyze, frame by frame, the movement of humans and other animals. As the Australian podiatric blogger Cameron Kippen, a self-described "retired academic, general bum, freelance writer, blogger and broadcaster" known as Toeslayer to his fans, notes, "An early pioneer of human movement was the comedian Charlie Chaplin who filmed many of his sequences backwards then ran them forward to accentuate the movement and expression."[1]

The big toe is so important to your movement and balance that a replacement for a lost toe appears to have been among the first prosthetics. Two such devices are still extant, one at the Egyptian Galleries of the British Museum in London, another at the Egyptian Museum in Cairo. The one in London, discovered at Thebes, is known as the *Greville Chester toe* after Reverend Greville Chester who acquired it for the museum 1881. The toe is made of linen-based papier-mâché covered with plaster tinted to look like skin and designed to fit a right foot. As Jacqueline Finch of the University of Manchester's KNH Centre for Biomedical Egyptology reported in *The Lancet* in 2011, a three-year examination of the artificial toe conducted by the British Museum between 1989 and 1992 concluded that "[b]ased on the 'spin characteristics of the linen' the toe could be dated to a time prior to 600 BCE." The second Egyptian false big toe, Finch continued, was even older, with a more complex structure. This one, again from Thebes, "was fastened onto the right foot of the female owner, who was identified as Tabaketenmut from around 950—710 BCE and was the daughter of a priest. . . . the carver seems to have been conscious of the anatomy and function of the foot. The inclusion of a hinge perhaps was intended to mimic the flexion of the metatarsophalangeal joint. Deliberate chamfering on the front edge shows an attempt to avoid rubbing against the navicular bone on the top of the foot, and the underside of the toe is flat for stability." Finch found two volunteers "with similar amputation sites," asked them to try on replicas of the two ancient Egyptian artificial toes, and found that they performed "extremely well" for motion and balance.

Your big toe can be useful in another way. In the event of digital disaster, the short, blunt, two-bone, one-joint appendage may be a lifesaver for your hand because it (the toe) can be amputated from your foot and grafted on to replace a lost two-bone, one-joint thumb, thus restoring your grasp.[2] Transferring your big toe from your foot to your hand may bar you from serving in the American armed forces because the U.S. Department of Defense thinks that having fewer than ten toes leaves you less balanced and less able to run to or from battle. But given the choice between one less toe and one new opposable thumb, the finger that makes it possible to perform such vital tasks as writing with a pen or opening jars, who would hesitate?

There is, of course, a price to pay for all these benefits.

Despite its virtues, the big toe is a place where good food sometimes goes to do bad things, triggering the arthritic pain in the toe called *gout* and in the process linking the digit to etymology, nutrition, genetics, sociology, the medical hierarchy, the green pharmacy, food and drug laws, and last, but certainly not least, one really important war.

## Naming, not taming, the beast

A disease isn't really official until you give it a name. Until then, it is, in the words of the very old vaudeville joke, "that thing you had before that you have again."

When a new, presumably infectious, disease appeared in the early 1980s, some doctors referred to it as lymphadenopathy because it caused swollen lymph glands. It seemed to strike only gay men, so others called it gay cancer or GRID, short for gay-related immune deficiency, but when the acronym AIDS for acquired immunodeficiency syndrome was proposed at a Washington conference in the summer of 1982, it stuck. By fall, when the CDC began to include AIDS in its *Morbidity and Mortality Weekly Report* (*MMWR*), a weekly compilation of public health reports, recommendations, and announcements, the cause, nature, and treatment of the disease were still a mystery, but now at least a mystery with a proper name.

AIDS was named within a year or two. That sometimes excruciating pain in your toe? Not so fast.

In the beginning and for some time after that, gout was known as *podagra*, from the Greek words *pod* meaning *foot* and *agra* meaning *catch* or *trap*. Podagra was also the name of a minor goddess, the daughter of Dionysus (Bacchus), the god of wine, and Aphrodite (Venus), the goddess of love. Given the fact that peoples across the Mediterranean blamed the aching toe on too much wine and—victims being mostly male—too much dalliance with the ladies, Podagra was the perfect avatar.

The first written description of podagra is commonly thought to be in Egyptian hieroglyphics in the Ebers Papyrus (c. 1500 BCE), the general treatise on disease and remedies named for Georg Ebers, the German Egyptologist who found the document in Thebes in 1873. Today, the Ebers Papyrus resides in the Library of the University of Leipzig, its antiquity rivaled only by the Edwin Smith Papyrus, a text on surgical treatment whose date, depending on your source, is either 1600 BCE or 3000–2500 BCE. This second scroll was discovered by American Egyptologist Edwin Smith, by coincidence born in 1822, the year Egyptian hieroglyphics were deciphered by Jean-François Champollion, the French scholar considered to be the Father of Egyptology. Champollion, who used the Rosetta stone to crack the code, was once considered the first to have done the job, but in 2004, archeologists at University College London discovered that Arab scholars had been there, done that, more than 1,000 years earlier, well before Smith bought the document from traders—read *thieves*—in Luxor in 1862. The scroll is now at the New York Academy of Medicine in Manhattan.

For centuries, most conditions causing joint pain were called rheumatism, from the Latin and Greek word *rheum* meaning *flowing*. In Old English rheum meant a thin watery discharge from your eye or nose, and, in the words of the *Oxford English Dictionary*, rheumatism was a medical condition assumed to be caused by "a 'defluxion' [sudden disappearance] of rheum," perhaps because arthritic joints were perceived to turn dry and crackly when they lost their natural lubrication. Eventually, with more precise diagnosis and definition, the single illness, rheumatism, became the arthritides, a category of autoimmune miseries running from A (ankylosing spondylitis) through V (viral arthritis). All are conditions in which the body attacks itself, inflaming and destroying connective tissue, the fibrous material that connects and supports organs, bones, and joints.[3]

Nobody called gout "gout" until the middle of the thirteenth century when Ralph Bocking (1197–1258), a.k.a. Randolphus of Bocking, a Dominican monk and chaplain to Richard de Wyche (1197–1253), Bishop of Chichester, proposed the name.[4] Bocking's choice of the word gout, from the Latin word *gutta* meaning a *drop*, reflected a view of disease adopted from Hippocrates' theory of the four humours, the basis of the Pneumatic School, an ancient Greek and Roman school of medicine. The idea was that good health depended on vital air (*pneuma*) and a balance among the four humours—yellow bile, black bile, phlegm, and blood—drops of elements in the body whose relationship determined both personality and physical health. Each humor was linked to a season of the year, an element of the environment, a degree of temperature and moisture, and a psychological condition:

Yellow bile = summer, fire, hot and dry, characterized by a hot temper.

Black bile = autumn, earth, cold and dry, characterized by melancholy.

Phlegm = winter, water, cold and moist, characterized by a calmness.

Blood = spring, air, hot and moist, characterized by passion and optimism.

The idea was scientifically wrong, but like most of Hippocrates' theories, a giant step forward nonetheless. With "germs" a concept still centuries in the future, Hippocrates challenged the prevailing view of illness by proposing that a person's health depended on conditions—humours—in the body rather than on divine intervention. As for Randolphus' suggestion that an ache in the big toe was due to drops of something slipping from the blood into the joint, although primitive, the premise was a step in the right direction.

Leading first to the dinner table.

## Podiatric protein problems

Protein molecules are built of chains of amino acids. It was long assumed that the fewest number of amino acids required to build a protein molecule was forty to fifty, but in 2004, scientists at the National Institute of Advanced Industrial Science and Technology (AIST) in Japan were able to synthesize one containing only ten "amino acid residues," amino acid molecules that have joined another amino acid and lost one water molecule in the process.

Chemistry is a science of continuing wonder. The Japanese researchers made a very small protein. By comparison, Titin, a stretchable protein that acts like an elastic spring in cardiac and skeletal muscles, has 34,350 amino acids, 539,022 atoms, a chemical formula that reads C169,723 H270,464 N45 688, O52 243S912, and a chemical name 189,819 letters long, making it the longest word in the English language. It is not easy to find the complete name. The first sites I came across that promised full disclosure showed this:

Methionylthreonylthreonylglutaminylala . . .

ylglutaminylprolylleucylglutaminylsery . . .

serylthreonylalanylthreonylphenylalan . . .

ylglycylphenylalanylprolylvalylprolyl . . .

anylarginylaspartylglycylglutaminylva . . .

Assuming that the dots (. . .) meant, "Pick up next line here," I spent about two hours doing just that. When I was done, I ran the stats and found I had filled fifteen pages, but had fewer than half the required number of letters. I finally discovered what purports to be the correct name, this time with 189,819 individual letters that filled 73 pages on my computer, which is why I am not reproducing it in its entirety here. Is what I found the correct sequence? I cannot say with absolute certainty. For all I know, the site that printed it missed or misplaced a letter here or there. But if it isn't the real thing, it's sure a good imitation. And it certainly beats *antidisestablishmentarianism*, the leading contender for world's longest word when I was in grammar school and we gauged our intelligence by our ability to spell it.

When you digest proteins, including the titins in beef, fish, poultry, lamb, and pork muscle foods, their large molecules break apart. One by-product of this process is purines, nitrogen compounds in protein foods. The name, chosen by German chemist Emil Hermann Fischer (1838–1914), awarded the Noble Prize in chemistry in 1902 for his synthesis of purines and sugars, comes from the Latin word *purum* meaning *pure*.

To the chemist, a purine is a heterocyclic aromatic organic compound, consisting of a pyrimidine ring fused to an imidazole ring. To the layman, this translates as a compound comprising more than one kind of carbon-containing ring-shape molecule in a structure that includes one specific ring-shape crystalline molecule with nitrogen atoms stuck to another ring-shape molecule with nitrogen atoms. Some purines occur naturally in body cells; others enter your body with the proteins in your food. The purines in body cells are called endogenous purines; those made when you digest proteins are called exogenous purines.

Caffeine, the stimulant that makes coffee stimulating, is an exogenous purine first isolated in 1819 also by a German chemist, Friedleib Ferdinand Runge (1795–1867). Runges' other major achievement was the invention of paper chromatography, a process used to identify the chemicals in an unknown mixture by putting a drop of the mixture onto a piece of absorbent paper, adding a solvent, and then watching as the components in the mixture separate out at different speeds and migrate to different sites on the absorbent paper. Eventually, this produces a pattern called a *chromatogram*, which you can compare to a diagram of known chemicals and thus figure out what's in your unknown mixture.

After Runge scored his caffeine coup, other chemists working independently identified several stimulants in tea and chocolate and guarana without realizing that they had simply rediscovered caffeine. In 1840, the work of two chemists named Martins and Berthemot revealed that everything everyone had found was identical to the caffeine extracted from coffee beans. Alas, although Martins and Berthemot are mentioned in most histories of caffeine, their first names and where they worked seem to have been lost to history. Neither contemporary sources such as the first edition of the *United States Dispensatory* (1868) nor the mighty Google turn up anything other than the date, their last names and the "independent" nature of their research.

As you can see from the drawings below, the chemical structures of caffeine, theophylline (from tea), and theobromine (from chocolate) are similar to the purines adenine and guanine, two endogenous purines found in DNA/RNA that play a role in the synthesis of DNA and the bonding of one cell to another.

Like protein molecules, purine molecules break apart during digestion, a metabolic divorce that produces a waste product called uric acid. About 20 percent of the uric acid in our body comes from digesting proteins; the rest is made in our liver from adenine and guanine. Ordinarily, excess uric acid dissolves in our blood and is eliminated through our kidneys. But if our body makes much too much uric acid or doesn't eliminate the compound efficiently, blood levels of uric acid rise (a condition known as hyperuricemia), and the excess uric acid crystallizes into sharp needles of monosodium urate, particles that clump together as chalky white tophi. The first person to see the uric acid crystals in tophi was the Dutch microscopy pioneer and naturalist, Anton von Leeuwenhoek (1632–1723) in 1679. In 1776, pharmacological chemist Carl Wilhelm Scheele (1742–1786) isolated uric acid in urine and mineralogist Tobern Bergman (1735–1784) found uric acid crystals in bladder stones. They did not have to look far to find their material: Both Scheele and Bergman were Swedes working in the country that was home to the high-purine herring, to the sauna famed for its ability to soothe arthritic pain, and to Carolus Linnaeus (1741–1783), the professor of Natural History at Uppsala University whose own gout did not keep him from creating the genus/species system for naming and classifying living creatures.

Before anyone knew about proteins or purines or uric acid, it was obvious that a diet rich in certain foods such as meat (particularly organ meats) and seafood (particularly anchovies and herring) were linked to a higher risk of gout. In 1905, *Meals Medicinal*, a collection of "Curative Foods from the Cook, in Place of Drugs from the Chemist," by W.T. Fernie, M.D., of

London, advised the reader that "unlike beer, or any other malt liquor, [cider] acts as an antidote to gout," "natural" wine provides a "natural immunity," fortified wines "set gout going viciously in the system," apple juice "neutralizes acid products of digestion or gout," blackberries and mulberries are "particularly wholesome," and cabbage, a sort of panacea, cures "constipation and dysentery, headache, and lumbago; retention, and incontinence of urine, pains in the liver, and affections of the heart; colic, toothache, gout and deafness."

Modern research confirms the culpability of some, but not all high protein, high purine foods. As a general rule, those yielding 150–1,000 milligrams purine in a 100-gram (3/5 ounce) serving are considered high in purine. Moderate purine foods yield 50–150 milligrams per 100 grams; low purine foods, 0–50 milligrams per 100 grams. If you have gout and must watch your diet, your doctor or nutritionist is likely to hand you a food list that looks something like this:

> *High purine foods*: Game (quail, rabbit, venison), organ meats (brains, heart, kidney, liver, sweetbreads), fish (anchovies, herring, sardines), shellfish, beer and wine.
>
> *Moderate purine foods*: Poultry (other than game), red meat (beef, lamb), fish other than the high purine varieties, whole grain bread, cereal and pasta, legumes (beans, peas and peanuts).
>
> *Low purine foods:* Dairy products (butter, eggs, milk, yogurt), fruits, broccoli, cauliflower, mushrooms, spinach—most vegetables other than asparagus—nuts and spice.

White, brown, or "raw," sugars are also low purine foods, thus considered acceptable for people with gout. More complicated sweeteners turn out to be more complicated.

In 2004, investigators at Massachusetts General Hospital (Boston) released the results of a 12-year-long study tracking the dietary habits of nearly 50,000 men who were gout-free at the start. Yes, the men who ate lots of meat (particularly organ meats) were more likely to develop gout, but gorging on high-protein vegetables such as beans and peas had no such effect, and eating lots of high-protein dairy foods actually lowered the risk. Four years after that, researchers discovered a new culprit hiding on the grocery shelf: The fructose-sweetened beverages targeted by anti-obesity experts and the Mayor of New York City who, in 2012, proposed banning soda servings larger than 16 ounces in public venues such as sports stadiums, thus making himself seriously unpopular with a wide swath of the populace. In 2008, nutrition scientists at the Arthritis Research Centre of Canada/University of British Columbia (Vancouver) announced that data from their own 12-year study of more than 46,000 male volunteers showed a higher-than normal incidence of gout among men drinking soft drinks sweetened with this particular sugar. Compared with men who drank less than one fructose-sweetened soda a month, the risk for men drinking 5-to-6 sweetened sodas per week went up 30 percent; it was 45 percent higher for men drinking a soda a day and a whopping 85 percent higher for those whose daily diet included two or more sweetened sodas.

Diet soft drinks? No problem. Substituting fruit juices? Possible problem. "[F]ructose rich fruits and fruit juices," the study authors warned, "may also increase the risk." This is one case in which an apple a day won't keep the doctor away because apples, applesauce, apple juice, pears, dates, and watermelon are all high fructose foods, and too much fructose may reduce your body's ability to flush away the excess uric acid that turns into crystals that turn into gout.

Blame it on your genes.

As you know, the early Greeks suspected that it made good sense to pair animals with "good" characteristics to produce "good" offspring, and that in humans some abnormalities such as clubfoot ran in families. In 451 BCE, Hermodorus took Plato's theories of eugenics for humans to the Decemvirs in Rome who wrote into law a father's duty to dispose of infants with birth defects. Six hundred years after that, Aretaeus of Cappadocia (c. 130–200), a Greek physician practicing in Rome, identified and named *diabetes* with the Greek word for *siphon*, a reference to the diabetic's frequent urination. Then he came up with a word to describe the source of human abnormalities, *diathesis*, Greek for *tendency*. Aretaeus didn't say a diathesis caused a specific disability, disease, or medical condition; his very smart and very advanced idea was simply that having the tendency—what we may now call a genetic predisposition—made a person more susceptible to developing the problem.

In *The Book of Animals*, African Arab scholar Al-Jahiz (Abu Uthman Amr ibn Bahr al-Kinani al-Fuqaimi al-Basri, c. 776–869) described natural selection centuries before Darwin. Latin translations of his work were in the libraries of Linnaeus and of Jean-Baptiste Lamarck (1774–1829), the latter most famous for wrongly proposing that an acquired characteristic could be passed along to succeeding generations. In *Al-Tasrif* (*The Method of Medicine*), another medical and surgical encyclopedia, the Arab physician, surgeon, and chemist Abu al-Qasim Khalaf ibn al-Abbas Al-Zahrawi, a.k.a. Albucasis (936–1013) seems to have been the first to describe the hereditary nature of hemophilia.

But then, amazing as it seems, the study of genetics receded into the scientific background until 1865, when a Moravian monk named Gregor Mendel (1822–1884) delivered two lectures on his observations of the plants in his monastery garden to the Natural History Society of Brunn (now known as Brno, in the Czech Republic) and then published them in the *Verhandlungen des naturforschenden Vereins*, the *Proceedings* of the Natural History Society in Brünn. Mendel was ignored as Raymond Dart had been when he wrote about the Taung Child, and like Dart, he was eventually rediscovered and rewarded, at least intellectually. In May 1900 British zoologist William Bateson (1861–1926), having come across Mendel's original article, arrived at a meeting of the Royal Horticultural Society in London to cite Mendel in a lecture on "problems of heredity as a subject for horticultural investigation." A mere 59 years later, Jerome Lejeune identified Trisomy 21 as the cause of Down syndrome, the first conclusive link between a chromosomal aberration and a medical condition. Forty-one years after that, on July 1, 2000, the Human Genome Project released the first draft of the human genome; in 2003, the map of the 20,000–25,000 genes that characterize the human race was completed. In 2012, ENCODE (Encyclopedia of DNA Elements), a multi-institution federal follow-up to the Human Genome Project, discovered that bits of DNA once regarded as "junk" actually contain several million "switches" that play an important role in the behavior of body cells, tissues, and organs.

That brings us to a less earth-shaking, but important moment in 2009, one year after the fructose/soda study, when a team of scientists from the MRC Human Genetics Unit at Western General Hospital, Edinburgh, and other centers in Britain, Croatia, and Germany, zeroed in on a genetic link between fructose and a higher risk of gout. One in four people born with clubfoot has a relative with the same problem; so does one in every four people with gout. The gene linked to clubfoot is called Pitx1. According to the MRC-led team, the genetic culprit for gout appears to be SLC2A9 (solute carrier family 2, facilitated glucose transporter member 9). This gene ferries sugars, including fructose, and uric acid about the body; a variant (mutation) seems to inhibit the body's ability to wash uric acid out of the bloodstream, send it to the kidneys, and then out into the world.

Of course, gene or no gene, not every pain in the toe means you have to give up your diet soda because not every pain in the toe is true gout.

Pseudogout, known colloquially as *false gout* and formally as *calcium pyrophosphate deposition disease* (CPPD), is a form of arthritis that commonly hits at the knees, ankles, elbows, and wrists. Like true gout, it is triggered by the collection of sharp crystals in the joints, but in this case the crystals are calcium pyrophosphate not monosodium urate. The calcium crystals are most likely to occur in older adults, often after severe dehydration leads to chondrocalcinosis, the formation of calcium deposits in cartilage.

Modern food safety rules have eliminated another risk factor that can make a toe ache: lead. The silvery gray metal was among the first four to be discovered, either along with or after the Big Three: gold (c. 6000 BCE), copper (c. 4200 BCE), and silver (c. 4000 BCE). Unlike gold, copper, and silver, lead is not found alone in nature; it is locked into the sulfur and lead ores galena (lead sulfide) and anglesite (lead sulfate), as well as the carbon and lead ore cerussite (lead carbonate). This was not a problem; lead melts at the relatively low temperature of 327°C/620°F, so it is easy to extract.[5] What the Romans got when they extracted lead was a soft, easily molded, and non-corrosive material perfect for the pipes that carried water throughout Rome and its provinces, thus bequeathing to us the word *plumbing* from the Latin word for lead, *plumbum*, meaning soft metal; the chemical symbol for lead is Pb.

The Romans knew that handling lead could be hazardous to one's health, so all lead mining in Rome was done by slaves. But once the metal was mined and extracted, it was liberally sprinkled through everyday Roman life, from top to bottom. Lead leeched into the drinking, cooking, cleaning water that flowed smoothly through the City's perfectly engineered and perfectly poisonous lead pipes. Lead made its way into food via flakes sloughing off plates and drinking vessels, or as "sugar of lead," sweet-tasting lead acetate crystals added to cakes and candies and fruits and the otherwise sour wine served in gleaming gold cups. All this lead, as well as the infinitesimal particles floating in the air, made its way into the bodies of those who ate or drank the foods and liquids, or handled the lead dishes, bowls, and jars on or in which they were served or stored.

Inevitably, this wide exposure led to an epidemic of lead poisoning. Two important symptoms are dementia—Caligula and Nero are presumed to have been victims—and sterility. As one anonymous Roman poet quoted by the U.S. Environmental Protection Agency, no less, concluded, both effects were devastating to the Roman state:

> Hence gout and stone afflict the human race;
> Hence lazy jaundice with her saffron face;
> Palsy, with shaking head and tott'ring knees.
> And bloated dropsy, the staunch sot's disease;
> Consumption, pale, with keen but hollow eye,
> And sharpened feature, shew'd that death was nigh.
> The feeble offspring curse their crazy sires,
> And, tainted from his birth, the youth expires.

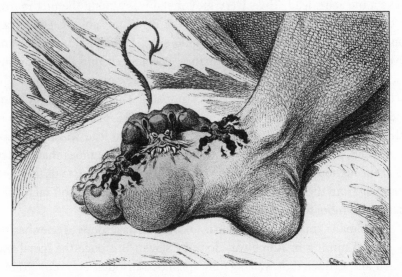

A third symptom of lead poisoning is the agonizing joint pain known as *saturnine gout*, after the planet Saturn that was early on believed to be made of lead. There was no remedy for any form of gout in ancient Rome and no remedy in 1898 when all the first edition of *The Merck Manual* could offer was sulphuric [sic] acid diluted in lemon juice to prevent the body's absorbing lead; alum, castor oil, magnesium sulphate, potassium iodide, and sulphur as laxatives; belladonna to calm the intestines; morphine and opium for the pain; and eggs and milk, presumably to soothe the stomach if not the foot.

## Power, prevention, and sexual prowess

Plain, false, or saturnine, the gout cloud did have a sliver of a silver lining. While writhing in pain, the sufferer could bask in the aura of wealth and power attested to by that inflamed toe. Thomas Sydenham (1624-1689), the seventeenth century British physician generally recognized as a founder of modern clinical medicine—the *Encyclopedia Britannica* calls him "the English Hippocrates"—and a gout sufferer himself, accurately reflected the prevailing wisdom when he wrote, "Gout, unlike any other disease, kills more rich men than poor, more wise men than simple."

In the language of classic medicine, gout was *morbus dominorum et dominus morborum*, the "disease of lords and lord of disease." The origins of this cozy relationship are described in the fable *Mr. Gout and the Spider*, a tale that's been told at least since the ninth century. One well-known version appears in *The Poore-Mans* [sic] *Plaster-Box*, a medical handbook published by Puritan minister Richard Hawes in 1634 for his country parishioners in Kentchurch, Herefordshire, in southwest Britain.

One evening, the story goes, as Mr. Gout and Mr. Spider were traveling the countryside, they began to look for shelter for the night. Gout went off to a poor man's home; Spider to a rich man's. When they met the next morning to compare experiences, each had the same complaint: "Mine was worse than yours." "As soon as I touched the leg of the poor man," said

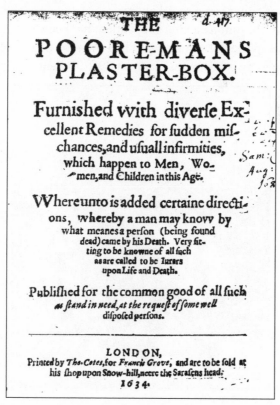

Gout, "he started moving around so wildly that I couldn't get any sleep." "That's nothing," said the Spider. "As soon as I began to build my web, the rich man's maid came and tore it down." The next night, changing places, the two discovered happy homes. The poor man left Spider's webs intact; the rich man pampered gout with soft pillows and tasty meals designed to sooth both the toe and the spirit. And that sealed the deal: gout for the rich, spiders for the poor.

Like spiders, rheumatism was also the poor man's fate, a situation nicely explained by the patrician Mr. Chester in Charles Dickens' *Barnaby Rudge*. "You will excuse her infirmities," he says of the maid employed to open doors. "If she were in a more elevated station of society she would be gouty. Being but a hewer of wood and drawer of water she is rheumatic." One literary lion rebelled at this convention. Daniel Defoe, both a failed businessman and a man afflicted with gout, wrote the hero's father into *Robinson Crusoe* as, yes, a poor man with gout, thus defying common wisdom, which remained common almost two hundred years later when *The London Times* commented that, "The common cold is well named—but the gout seems instantly to raise the patient's social status."

But Crusoe's father was a social anomaly.

Historically, the quintessential gout patient has been forty, fat, and filthy rich, or at least powerful. Shakespeare had the type down cold with John Falstaff, the "goodly portly man, i' faith, and a corpulent; of a cheerful look,

a pleasing eye, and a most noble carriage" in *Henry IV, Part 1.* By *Part II,* he was no longer so cheerful, invoking "A pox of this gout! or, a gout of this pox! for the one or the other plays the rogue with my great toe."

As you may have noticed, the gout sufferer has almost always been a *he.* In his *Aphorisms,* Hippocrates noted that "a woman does not take the gout, unless her menses be stopped" or she begins to behave like a man. During Nero's reign, the philosopher and statesman Lucius Annaeus Seneca, a.k.a. Seneca the Younger or just plain Seneca (c. 3 BCE–65 CE), fretted that although women were not ordinarily destined to lose their hair or suffer gout, "because of their vices, [they] have ceased to deserve the privileges of their sex; they have put off their womanly nature and are therefore condemned to suffer the diseases of men." But as noted above, what looked like the wages of sin was really Toxicology 101: Like dementia and sterility, hair loss and joint pain are symptoms of lead poisoning.

Today, the ratio of men to women with gout remains about 4:1 in favor of (against?) men. Until recently, those were assumed to be primarily Caucasian men, but modern statistics refute this. In November 2011, at the Association of Rheumatology Health Professionals Annual Scientific Meeting in Atlanta, a team of scientists from Johns Hopkins reported that after combing through the records of 15,792 men and women in the Atherosclerosis Risk in Communities (ARIC) Study and accounting for all the usual variables—gender, age, weight as measured by the Body Mass Index (BMI), diet (protein, organ meat, shellfish, alcohol), chronic conditions (hypertension, diabetes, kidney function), income, education, smoking, and body levels of uric acid—they found a 50 percent higher risk of gout among African American men. And women.

Gout was also believed to convey sexual prowess. "Eunuchs do not take the gout, nor become bald," Hippocrates wrote. "A young man does not take the gout until he indulges in coition." Remember, gout was originally called *podagra* after the daughter of the Greek god of wine and the goddess of love. Counterintuitive as it might be, some even considered gout an aphrodisiac. In his essay, "On Cripples," Michel Eyquem de Montaigne (1533–1592) wrote that when the legs "do not receive the food that is their due . . . the genital parts . . . are fuller, better nourished, and more vigorous. Or else that, since this defect prevents exercise, those who are tainted by it dissipate their strength less and come more entire to the sports of Venus." More to the point, others suggested, just lying around on your back doing nothing much tends to turn your mind to thoughts of a baser nature— which you might not be able to complete owing to the pain in your big toe.

Finally, a kind of it-could-be-worse mentality credited the disease of lords with the ability to ward off less noble, but potentially more disastrous illness. The eighteenth century literati were fulsome in their praise of gout as preventive medicine. In a poem to celebrate his friend Rebecca Dingley's birthday ("Bec's Birthday" [1726]), Jonathan Swift (1667–1745) wrote of "The gout that will prolong his days." British historian, playwright, and novelist Horace Walpole (1717–1797) believed that gout "prevents other illness and prolongs life. Could I cure the gout, should not I have a fever, a palsy or an apoplexy?" And Ben Franklin (1706–1790), author of *Poor Richard's Almanack* and a man accustomed to publishing sensible advice, described his own straight-forward conversation with his disease in *Dialogue Between Franklin and The Gout* (1780):

> GOUT. Well, then, to my office; it should not be forgotten that I am your physician. There.
>
> FRANKLIN. Ohhh! What a devil of a physician!
>
> GOUT. How ungrateful you are to say so! Is it not I who, in the character of your physician, have saved you from the palsy, dropsy, and apoplexy? one or other of which would have done for you long ago, but for me.
>
> FRANKLIN. . . . [O]ne had better die than be cured so dolefully. Permit me just to hint, that I have also not been unfriendly to you. I never feed physician or quack of any kind, to enter the list against you; if then you do not leave me to my repose, it may be said you are ungrateful too.
>
> GOUT. I can scarcely acknowledge that as any objection. As to quacks, I despise them; they may kill you indeed, but cannot injure me. And, as to regular physicians, they are at last convinced that the gout, in such a subject as you are, is no disease, but a remedy; and wherefore cure a remedy?—but to our business,—there.

Defoe, Swift, Walpole, and Franklin were only four of gout's middle-aged male victims. The list is so long and impressive that you just have to wonder how strongly their gout influenced their public behavior.

Did the pain of that big toe drive Alexander and Charlemagne into battle, make Henry VIII so cranky he just had to change wives every few years, embitter Benedict Arnold, and turn Karl Marx into an opponent of the upper class into which his ailment supposedly placed him? If Martin Luther hit his thumb when nailing those 95 theses to the door of the Wittenberg Castle Church, was that a minor inconvenience compared to his gout? Did Galileo challenge the map of the universe and invite theological condemnation just to take his mind off his foot? Did concentrating on sounds he heard only in his head distract Beethoven from his own throbbing toe? Did the shock of that apple falling on Isaac Newton's head turn his attention from physics (as in remedy) to physics (as in gravity), an event often assumed to be apocryphal, but actually to be found in a contemporary biography, *Memoirs of Sir Isaac Newton's Life* (1752), by fellow gout-sufferer William Stukeley. The book, published twenty-five years after Newton's death, is online today at the website of the Royal Society in Britain where you can read Stukeley's account of how "[a]fter dinner, the weather being warm," he and Newton "went into the garden and drank thea [sic], under the shade of some apple trees. He told me, he was just in the same situation, as when formerly, the notion of gravitation came into his mind. It was occasion'd by the fall of an apple, as he sat in contemplative mood. Why should that apple always descend perpendicularly to the ground, thought he to himself . . ."

And when the pain of gout inhibited motion, how long would patients have had to wait for the invention of the wheelchair had King Philip II of Spain's painful toe not required the first relatively modern documented moving chair, shown here in a drawing thought to be dated sometime around 1595, three years before the king and his toe exited the world.

Finally, what role did the misery of his own imperfect, gouty toe play in Leonardo's fascination with anatomy and the perfect *Vitruvian Man*?

We will never know. But we do know this: Gout played a rarely reported, highly personal role in the lives of five men in America, England, and France, who managed or tried to contain the emancipation of the United States from Great Britain.

## Gout's white shoe fraternity

Human beings are contentious by nature, especially when gathered together into families, tribes, or nations. Push them a bit, and they will take up arms for all kinds of causes, reasonable and otherwise: A (married) Spartan queen's elopement with a Trojan prince; a French king's refusal to grow back the beard shaved after his return from the Crusades, leading his wife to divorce him, marry the King of England and demand her dowry back, which King #1 refused to pay, thus leading King #2 to declare war on King #1; a British seaman's ear amputated with one slice of a Spanish commander's sword; the sinking of an American warship in Havana harbor, or the assassination of an Austrian Grand Duke.

This list of conflicts—the Trojan War (1193–1183 BCE), the War of the Whiskers plus the War of the Austrian Succession (1152–1453), the War of Jenkins' Ear (1740–1748), the Spanish–American War (1898–1901), and World War I (1914–1918)—would be incomplete without the "War Hastened by A Gout Attack," better known as the American Revolution.

William Pitt, the Elder, 1st Earl of Chatham, Viscount Pitt of Burton-Pynsent (1708–1788), twice prime minister of Great Britain and known fondly as the "Great Commoner" for his service in the House of Commons until he was lifted not-quite-happily into the House of Lords in 1766, did not like the French. Disagreements with various kings and other important personages in England kept Pitt from being named Prime Minister until 1766 when he served for two years and seventy-six days, but he was able nonetheless to wield his razor sharp wit and tongue to convince his colleagues to underwrite Prussian engagements against the French in Europe and to send the British Navy to harry the French in ships off the coast of France, as well as in such outposts of French exploration and conquest as the West Indies and Africa during the Seven Years War. His policies enabled Britain to consolidate power in North America and on the Indian subcontinent, hold on to bases in the Mediterranean, and acquire territory in the previously French-dominated Africa and West Indies. Then, having turned England from a kingdom into an Empire, Pitt was determined to hold it together for as long as possible, a situation which required not only outwitting the French, but also keeping the American Colonies relatively peaceful.

It wasn't easy. In 1765, when Pitt was absent due to a disabling bout of gout, Parliament passed the infamous Stamp Act commanding the Colonies to pay taxes to the British government to make up the cost of defending them (the Colonies) from attack by the French. As soon as he could, Pitt more or less hobbled back to Parliament, his gout in momentary retreat, and convinced his allies to repeal the Act. "The Americans are the sons, not the bastards, of England," he announced. "As subjects, they are entitled to the right of common representation and cannot be bound to pay taxes without their consent." But the next time Pitt's toe laid him low, Parliament repealed his repeal, and in 1773 passed the Tea Act, levying duties on tea, which the colonists famously tossed overboard into Boston harbor.

Clearly, Pitt's on-again, off-again gout attacks did not trigger either the colonists' drive for separation from the country they had left and so many of them had left behind or the British Parliament's equally strong intention to keep the Colonies subservient within the Empire. But Robert Burns was right: The best schemes often go awry—especially when, as in this case, the leading players are indisposed. Pitt never supported independence for the Americans because he feared French domination of North America, but he was sympathetic to their situation even after the Revolution began. "If I were an American," he said in November 1777, "as I am an Englishman, while a foreign troop was landed in my country, I never would lay down my arms—never! never! never!"

That being true, if Parliament had stuck with the Pitt-driven repeal of the Stamp Act in 1765 and not raised taxes on tea, could there have been large or small alterations along the way to full independence? Probably not, but even as late as 1775, Pitt was not ready to give up. After consultation with Benjamin Franklin, he introduced a compromise bill to maintain the legislative authority of Parliament over its Colonies while granting the Continental Congress in Philadelphia the responsibility of setting each colony's taxes to the British treasury. His proposal was turned down. One year later, as he had feared, Britain and the Colonies were at war. Luckily for the colonists, four equally eminent men who shared the American desire for independence—and one painful problem—were at work to make sure there would be arms sufficient to accommodate the American patriotism.

Franklin, Jefferson, de Vergennes, and Hancock. Say the names quickly and it sounds suspiciously like the list of partners in a white shoe law firm.[6] Or the clique of really cool guys at an exclusive boys' school, which it sort of was.

By age nine, Thomas Jefferson (1743–1826) was studying Latin, Greek, French, and "natural history" at schools run by ministers near his birthplace in Albermarle County, Virginia; and he graduated from William & Mary in 1752 at nineteen. In 1819, after leaving the Presidency, he founded the University of Virginia, firmly in the tradition of secular classical education. Hancock (1737–1793) was the first man to sign the Declaration of Independence, in a script so much larger than everyone else's on the page that his name became a synonym for a signature. He had attended the Boston Latin School, established by Puritan settlers in 1635 as the first public school in America, and graduated from Harvard College, class of 1774. In the manner of mannered French noblemen, Charles Gravier, Comte de Vergennes (1717–1787), foreign minister for Louis XVI, was educated by Jesuits in Dijon.

The odd man out might appear to have been Franklin, the son of a Puritan Boston candle maker. Like Hancock, Franklin attended Boston Latin, but sinking family finances forced him to drop out a few months after he entered. His formal education ended when he was ten; after that he taught himself. Later, Franklin won his key to the cool kids' club with two honorary master's degrees (Harvard, 1753; William & Mary, 1756) and two honorary doctorates (University of St. Andrews, 1759; Oxford, 1762). By 1856, when Richard Saltonstall Greenough's 8-foot-tall bronze statue of Franklin, the city's first such portrait of a public person, was set in place at the site of the original Boston Latin School and the Old City Hall, Franklin's adult achievements had clearly eclipsed his boyhood poverty.

This quartet shared not only a love for politics, but also the torture of their aching toes. It is easy to imagine them sitting around a fire in the evening, comparing aches and pains (actually it would have to be imagined because while Franklin, Jefferson, and de Vergennes were making nice in France, Hancock was back home in the Colonies).

Franklin, the only man to have signed all four documents that created the United States—the Declaration of Independence (1776); the Treaty of Alliance, Amity, and Commerce with France (1778); the Treaty of Peace between England, France, and the United States (1782); and the Constitution (1787)—was subject to attacks of gout so painful that, unable to walk, he was delivered to the Constitutional Convention in a sedan chair hoisted aloft by four hardy prisoners from the Walnut Street jail in Philadelphia. In 1787, Jefferson wrote from Paris to John Jay, the future Chief Justice of the United States, that the Comte de Vergennes was "very seriously ill. Nature seems struggling to decide his disease into a gout. A swelled foot, at present, gives us hope of this issue." As for his own gout, Jefferson dealt with that by soaking his feet in cold water each morning; there is no record of whether he followed the advice of various Old Wives to add willow bark (American Indian), stinging nettles (Swiss), salt (Russian), cider vinegar or pine extract (New England) to the water and no record either of the remedies' effectiveness. Meanwhile, at home in Massachusetts, Hancock cannily used the threat of his gout to political advantage when as Governor of Massachusetts, he took his time showing up to sign the Constitution, citing a convenient attack of toe throb. His tactical delay forced the Federalists to dance an offer of the vice presidency in front of him and to promise the addition of a Bill of Rights, thus prompting a most fortuitous recovery; in February 1788, Massachusetts became the sixth colony to ratify the Constitution. That same month, with de Vergennes' enthusiastic support, New America and Old France signed Treaties of Amity and Commerce granting military aid to the Colonies and mutual most-favored-nation status.

Gout, on the other hand, had long ago been accorded most favored disease status, miserable to be sure, but as we have seen, also surprising proof of the sufferer's virtue and medical and libidinous good fortune.

But how to soothe the unfortunate pain?

## Cutting corns and cultivating customers

The search for an effective way to ease an aching human foot is probably as old as the human foot itself. We have no studies from the caves, but drawn and written records tell us that the Chinese practiced chi (qi) foot massage, an ancestor of modern reflexology, more than 5000 years ago. As early as 2400 BCE, the Egyptians carved a bas relief of a foot operation on to Ankhmahor's Tomb, a structure sometimes called The Physician's Tomb. Although Ankhmahor was not a doctor but a politician whose proper title was *Vizier, First under the King, Overseer of the Great House*, the tomb was decorated with scenes of medical procedures. Hippocrates prescribed a vinegar bath to soften corns and calluses, a remedy that along with lemon juice or, in a pinch, urinating on the corn or even into your shoes, remained popular for centuries. Type "natural remedies for corns" into the search bar on your computer, and lemon juice and vinegar are still there, along with chamomile tea and a modern bath of aspirin dissolved in water. What all these remedies have in common is that each is an acidic solution that accelerates the body's natural flaking of the top layers of skin, softening a corn so it may be peeled or scraped away.

For hundreds of years, foot care peddled along in the backwaters of medicine, practiced by people known as "corn cutters." Kings and Queens had their own podiatric practitioners at court; lesser mortals found theirs on the street. In Britain, anyone might hail one of those who advertised their wares and work in the songs known as "The Cries of London" promising

Here's fine herrings, eight a groat;*
Hot codlines pies and tarts.
New mackerel I have to sell.
Come buy my Wellfeat & Oysters, ho!
Come buy my whitings fine and new.
Wives, shall I mend your husband's horns?
I'll grind your knives to please your wives,
and very nicely cut your corns.

---

* Four pence; for comparison, according to the "Time Traveller's guide to Tudor England" (http://forums.canadiancontent.net/history/49884-time-travellers-guide-tudor-england.html), at the time a diet staple for the poor was a "halfpenny loaf of bread, which feeds two people."

Despite the lowly status of the corn cutter, problematic feet often made their way into art and literature, such as this passage from *Romeo and Juliet*, Act I, Scene 4. Or Act I, Scene 5, depending on the source. *Bartlett's Familiar Quotations* (16th ed., 1992) has five scenes in Act I; *The Yale Shakespeare* (1993) has four. Either way, Juliet's father, Lord Capulet, hosts a dinner dance where Romeo, seeing Juliet for the first times, rhapsodizes: "Did my heart love till now? forswear it, Sight! For I ne'er saw true beauty till this night," which surely beats the modern adolescent accolade, "Awesome!" Back at the dance floor, Capulet encourages participation in the festivities by slyly suggesting that ladies who decline to dance have un-romantically aching feet:

Welcome, gentlemen! Ladies that have their toes
Ah, my mistresses! Which of you all
Unplagued with corns will walk a bout with you.
Will now deny to dance? She that makes dainty,
She, I'll swear, hath corns. Am I come near ye now?

Treating feet eventually became more respectable, but it was a slow trek up from the bottom. The number of corn cutters in England increased after William of Orange deposed the Catholic James II and ascended to the British throne in 1689, bringing with him his own squad from the Netherlands. The group included John Hardman, a man so well regarded that not one, but two portraits of him hung in the British National Gallery. A second corn cutter, one Mrs. Seymour Hill, was so widely known in London and her clientele so fashionable that Charles Dickens is said to have chosen her as the model for the dwarf beautician, gossipy Miss Moucher in *David*

*Copperfield*. When the novel appeared, Mrs. Hill reputedly buttonholed the author and upbraided him for satirizing her but was mollified by Dickens' explanation that her character ended up a heroine.

Hartman and Hill aside, cutting corns and shaving calluses were still the low end of the healing arts. In 1800, *Kelly's London Directory*, a sort of pre-telephone telephone book that listed, in alphabetical order, "persons holding situations under the crown in the Bank of England, the various Law, City, and all other public offices," local officials such as mayors, heads of households, commercial traders, judges and lawyers, members of Parliament, banks and the various forms of transportation." There was only one foot specialist.

Then came the revolution. It started with the introduction of a fancier name: *chiropodist*, which etymologists consider either a smoother version of the earlier *chirurgapodist*, from the Latin *chirurgia* for *surgeon*, or a combination of *chiro*, the Greek word for *hand* or *foot*, plus the Latin *pod* meaning *foot*. The word appears to have been invented by a Londoner named David Low. Looking to enhance his reputation, he appropriated as his own a book called *L'Art de Soigner les Pieds* (*The Art of Caring for the Feet* [1781]), actually written by Louis XVI's personal corn cutter, Nicholas-Laurent LaForest. Low changed the title to *Chiropodologia*, but was soon exposed as a plagiarist; LaForest's original book remained the standard treatise on foot care. In 2012, Bauman Rare Books, a New York antiquarian bookseller, offered for sale at $16,500 a copy with Marie Antoinette's very own gilt crest on the cover, "Second edition (first published in 1781) of this classic podiatric work, written by the chiropodist of Louis XVI and the royal family, with the gilt arms of Marie Antoinette, beautifully bound. . . . Small octavo, contemporary full brown morocco, elaborately gilt-stamped spine and covers bearing the armorial crest of Marie Antoinette, marbled endpapers, all edges gilt. $16,500."

The word *chiropodist* was eventually replaced by the modern *podiatrist* during the first decades of the twentieth century, a period that also brought the introduction of dedicated professional training; a professional degree, Doctor of Podiatric Medicine; and several professional societies (some of which still use the name "chiropodists") along with the requisite professional journals. Today, like dentistry, podiatry remains a separate branch of the health sciences with its own schools. And like dentists, who with further training may become oral and maxillofacial surgeons, treating complicated injuries, abnormalities, and diseases of the head, neck, and face, podiatrists may choose to specialize in various fields, in this case primary care, pediatrics, geriatrics, diabetic wound care, or foot and ankle surgery.

## Safe, effective—and expensive

Before there were doctors and drugs, there were herbalists and the green pharmacy, the plants that yielded effective medicines whose benefits we still enjoy. The dental analgesic eugenol comes from clove oil, the painkiller salicylic acid (our modern aspirin) from the bark of the willow tree, and the painkillers codeine and morphine from the pretty red poppy. Celery seeds and dandelions produce diuretic oils. The anti-hypertensive, anti-arrhythmic digitoxin (Digitalis) comes from foxglove; the sleep-aid Valerin (valeric acid) from teas made with valerian roots and stems; the anti-malarial quinine from the bark of the South American cinchona tree

Another medicinal plant, meadow saffron (*Colchicum autumnale*), also known as the autumn crocus, has been part of the natural pharmacopoeia for as long as there have been written medical records. The plant, native to Eastern Europe and Asia, is named for Colchis, the ancient province east of the Black Sea where legend says the Greek hero Jason conspired with the sorceress Medea to steal from her father the legendary Golden Fleece that would win Jason back his rightful kingdom. The Ebers papyrus describes meadow saffron as soothing to the joints. The first person known to prescribe it specifically to relieve the pain of gout was Alexander of Tralles, a sixth century Greek physician practicing, as most of them seem to have done, in Rome. Alexander was well known and respected as the author of *Letter on Intestinal Worms,* the first treatise on parasitology, and the *Twelve Books on Medicine*, available in Latin, Greek, and Arabic editions. Tralles' treatment did ease his patients' pain, but meadow saffron was a risky remedy. The plant is poisonous, a purgative whose adverse effects may include nausea, abdominal pain, bloody vomit and diarrhea, high blood pressure, and respiratory failure followed by death; Greek slaves are said to have used it to commit suicide. Thomas Sydenham, who had created laudanum, the opium and sherry mixture whose addictive soothing qualities ensnared such luminaries as Mary Todd Lincoln, Samuel Coleridge, and Franklin, considered the colchicum plant too dangerous for human use. But like modern advocates for medical marijuana, gout victims begged to disagree, and, Sydenham's warnings were ignored. The herb, well regarded on the Continent and listed in the *London Pharmacopeia* starting in 1616, eight years before Sydenham was born, remained the remedy of choice.

Traveling to London on Colonial business in 1757 and 1764, Franklin spent much time in the coffee houses of London with the members of "The Club Of Honest Whigs" where, as Samuel Johnson's biographer James Boswell recalled, "Some of us smoke a pipe, conversation goes on pretty formally, sometimes sensibly and sometimes furiously: At nine there is a sideboard with Welsh rabbits and apple-puffs, porter and beer." Many of "us," it should be noted, suffered from gout and were treated with remedies brewed from the colchicum plant. Franklin was neither a fool nor a fan of suffering; after one trip abroad, he returned to America with a supply of colchicum plants for himself and his fellow gout sufferers.

Colchicum wasn't the only plant Franklin carried back across the Atlantic. The Chinese tallow tree is a horticultural wonder. It thrives in sun or shade, flood or drought, all the while producing waxy berries that can be used to make candles or burned as fuel. "'Tis a most useful plant," Franklin wrote to Noble Wimberly Jones (c. 1723–1805), a British ex-pat whose support of the French and American revolutions led him to immigrate to America where he settled in Georgia, became a leader of the Whigs in Georgia, eventually serving in the Continental Congress. Unfortunately, the seeds Franklin gave to Jones were, as predicted, extraordinarily hardy. Today, the tallow tree, also known as the popcorn tree, has spread throughout the South and into the Gulf states, crowding out native plants such as the American holly. Kerry Crisley of the Nature Conservancy, writing on the *Cool Green Science*, a.k.a. "the conservation blog of the Nature Conservancy," called importing the seeds the "one dumb thing that Benjamin Franklin did."

Franklin's other import, however, was doing well while doing good. With a few tweaks here and there.

In 1820, two French chemists, Pierre Joseph Pelletier (1788–1842) and Joseph Bienaimé Caventou (1795–1877), isolated the active anti-gout ingredient in meadow saffron seeds, flowers, and corms (the corm is the round underground stem in which the plant stores its food). Thirteen years later, the German chemist P.L. Geiger purified and named colchicine. After that, the goal was to identify a safe and effective dose, a job that required the sacrifice of several unlucky kittens under perfectly horrible conditions. Tracing the history of colchicine, the 1868 edition of *The Dispensatory of the United States*, reports that one-tenth of a grain of the substance extracted from meadow saffron seeds triggered exactly the adverse effects in an eight-week old kitten that Sydenham thought too dangerous for humans. "Violent purging and vomiting were produced, with apparently severe pain and convulsions and the animal died at the end of twelve hours," while "[a] kitten somewhat younger was destroyed in ten minutes by only the twentieth of a grain." As for humans, the *Dispensatory* reports that "in doses sufficiently large to affect the system, [colchicine] gives rise to more or less disorders of the stomach or bowels, and sometimes occasionally acts as a diuretic and expectorant; and a case is on record of violent salivation . . . while it somewhat diminishes the action of the heart. In an overdose, it may produce dangerous and even fatal effects." In short, exactly the consequences that made colchicine the poison of choice for Greek slaves and convinced Thomas Sydenham that it was too dangerous to be used as medicine. Nonetheless, relatively high doses were the rule in the United States as well as Europe, the standard prescription being to "take until diarrhea occurs." Not until 2009, 103 years after Congress passed the first Pure Food and Drugs Act in 1906, did the government begin to address the problems caused by what was considered a "normal" dose of the gout remedy from the meadow saffron plant.

The original Pure Food and Drugs Act arrived in 1906, one year after Upton Sinclair's novel *The Jungle* was published in serial form in the weekly political newspaper *Appeal to Reason*. The novel was an immediate sensation. "I aimed at the public's heart," Sinclair (1878-1968) said, "and by accident I hit it in the stomach." One prominent stomach belonged to President Theodore Roosevelt who, having read Sinclair's revolting depiction of the Chicago slaughter houses where workers, animals, and food itself were abused, pressed strongly for passage of the food law.

The 1906 law was administered by the Bureau of Chemistry of the U.S. Department of Agriculture. In 1927, Congress created the Food, Drug, and Insecticide administration, within the U.S. Department of Agriculture. Three years later, this division became the U.S. Food and Drug administration. In 1933, when it began an overdue drive to update the 1906 law, the agency faced an uphill battle with manufacturers on all sides, but tragedy led to remedy, first in cosmetics and second in drugs.

As FDA began its work, at least 12 women were blinded and one died after using Lash Lure, an eyelash dye sold as "permanent mascara." The culprit in the product was p-phenylenediamine, a coal tar dye that caused blisters, infections, and ulcers on the eyes, lids, and face of customers (all modern hair dyes carry a warning to avoid the eyes). The scandal not only reinforced the need for new laws; it was a bonanza for Maybelline which introduced cake eyelash and brow darkener in 1917. The coloring, packaged as a cake and applied with a moistened brush, was re-christened "mascara" and advertised in 1934 as "genuine, harmless Maybelline. Non-smarting, tear-proof Maybelline is NOT a DYE, but a pure and highly refined mascara for instantly darkening and beautifying the eyelashes."

Four years later, in the fall of 1937, the motivating factor was a drug disaster—more than one-hundred Americans in fifteen states died after drinking Elixir Sulfanilamide, a raspberry-flavored antibiotic syrup manufactured by the S. E. Massengill Pharmaceutical Company. The antibiotic had been dissolved in diethylene glycol (DEG), a sweet-tasting syrupy liquid found in antifreeze. The product was tested for flavor, but not for toxicity; although the subject was under consideration, at the time, the law required testing for purity, but not for safety. Unfortunately, diethylene glycol is a poison, and the Elixir appeared to contain more than the potentially lethal amount per dose. Testing food and drugs for safety was implicit if not explicit in FDA's name and on the agenda even before the Elixir event. Now, the Lash Lure injuries prompted the inclusion of cosmetics. In 1938, Congress finally passed the Food, Drug, and Cosmetic (FD&C) Act with new safety provisions for all three categories of consumer products. Then, in 1962, in the wake of the thalidomide disaster, a new set of amendments named for Senator Estes Kefauver (1903–1963) and Representative Oren Harris (1903–1997) required manufacturers to prove their drug not only safe, but also effective before it could be approved for marketing.

All drugs have side effects (usual and expected consequences); some have adverse effects (rare and unexpected consequences). When patients ask about possible problems with a new medicine, doctors are likely to respond that if aspirin were introduced today, given its side effects and adverse effects, it might never win FDA approval.

For example, taking aspirin may actually double your risk of gastric bleeding, and as a June 2011 editorial in *American Family Physician* notes, it has been clear for nearly twenty years that enteric-coated or buffered aspirin does not necessarily decrease the risk of gastrointestinal problems. As we grow older and the stomach lining thins, the risk is even higher, which is why the prescription for long-term aspirin therapy to reduce the risk of heart attack is a daily "baby aspirin," 75 or 81 milligrams rather than the standard 325 milligrams tablet. True, aspirin's ability to "thin" blood, that is, to prevent platelets from sticking together to build artery-clogging clots, certainly reduces the risk of heart attack and stroke in some patients, but which patients and by how much?

How do you let this new drug, aspirin, go on sale if you can't answer those questions? And if the drug, aspirin, has been used for decades or even centuries without the requisite tests, how do you allow it to stay on the market?

The answer to the first question is, you don't. The answer to the second is that you may choose to leave it in place and catch up when you can, perhaps when a new use is proposed.

Obviously, FDA was not around to evaluate colchicine when Alexander of Tralles first prescribed it, or to say *yes* or *no* to Thomas Sydenham's argument that it might be too dangerous for human beings. The 1938 Food, Drug, and Cosmetic Act did require all *new* drugs to be approved by FDA for safety, but it also permitted drugs already on the market, such as generic colchicine, to remain available. In 1976, when Merck & Company created Col-Probenecid, a pill containing colchicine and the drug probenecid, which lowers uric acid, FDA did test and approve the combination as a "new" drug. But it was considered unlikely that any drug company would invest the millions of dollars required to prove safety and effectiveness for plain colchicine, a natural substance that had been used for centuries and was available to anyone who wanted to go out, gather meadow saffron, and extract the active ingredient for which an effective dose had already been established.

Unlikely, that is, until one company did. In 2007, URL Pharma, a division of Takeda Pharmaceuticals U.S.A., in turn a subsidiary of Takeda Pharmaceuticals Company Limited of Japan, began testing a lower-dose colchicine product called Colcrys and discovered that it worked as well as the customary higher-dose, longer-term regimen. This was, as *The New England Journal of Medicine* explained, "technically a new indication for the drug." In 1984, Congress had passed the Waxman-Hatch Act (1984) that contained provisions to increase the availability of less-expensive generic drugs. Under this law, FDA was required to award URL Pharma three years of "market exclusivity" for its "new" generic colchicine, meaning the company would be the only one permitted to sell the drug.

URL Pharma also proposed to market Colcrys as a treatment for familial Mediterranean fever (FMF), a genetic disorder that worldwide affects perhaps 100,000 people of Armenian, Arabic, Turkish, and Jewish ancestry. Although colchicine was already known to help control the pain and fever of FMF, URL Pharma provided more safety information based on the trials for the low-dose colchicine gout product. The company's intention to propose Colcrys as treatment for FMF put the medicine under the protection of yet another law: The Orphan Drug Act. This law, passed in 1983, encourages research on drugs for diseases such a FMF that affect relatively small numbers of people; therefore, they do not guarantee the immense profits that will accrue to, say, a new antibiotic that can treat millions. The incentive in the law is a second guarantee of exclusivity, this time for seven years.

In July 2009, FDA officially announced what doctors and patients had known since Alexander's day: Colchicine effectively treats acute flares of gouty arthritis.

No surprise there.

What did surprise many patients and physicians was the FDA-approved double dose of exclusivity, which immediately produced one of those moments best described as proof of the Doctrine of Unintended Consequences. As the *New England Journal of Medicine* reported, "Once FDA approved Colcrys, the manufacturer brought a lawsuit seeking to remove any other versions of colchicine from the market and raised the price by a factor of more than fifty, from $0.09 per pill to $4.85 per pill [and the average 23-day prescription from $6.72 to $185.53]. According to the Centers for Medicare and Medicaid Services, state Medicaid programs filled about 100,000 prescriptions of colchicine in 2007 and paid approximately $1 million for the drug. Use of the new brand-name colchicine could add as much as $50 million per year to these insurance ad programs' budgets at a time when they are addressing the rising costs of health care by reducing some services or raising eligibility thresholds."

There may be another such issue on the horizon. Colchicine interrupts mitosis, the process by which cells divide their chromosomes to reproduce themselves. As a result, some of the resulting new cells may have more than the normal amount of genetic material. This is useful for breeding new and unusual plants, but not so good for human beings—unless the cells whose division colchicine interrupts are malignant.

As early as 1939, colchicine had been put through trials to evaluate its effects on patients suffering from leukemia; today is it sometimes used together with more conventional anti-tumor medication. In September 2011, researchers from the Institute for Cancer Therapeutics (ICT) at the University of Bradford (England) reported that a "smart bomb" using colchicine, chemically modified to make it inactive in the body until it reaches the tumor, had been able to slow the growth of five types of cancer—breast, colon, lung, sarcoma, and prostate—in laboratory mice without appearing to cause adverse effects. In some cases, it killed the tumor outright by destroying the blood vessels that keep the malignancy alive. And it worked fast. In one study, 50 percent of the mice were in complete remission, no tumor evident, after only one dose. If such results continue to be replicated in animal studies and the drug is approved for trials on humans and the results are positive, there is the possibility of a new approval (and more exclusivity) for even a very old drug.

But not just *any* very old drug. Colchicine, the meadow saffron plant's remedy for the king of diseases may now join the anti-tumor drugs vinblastine (Velban) and vincristine (Oncovin) from the Madagascar periwinkle in the biggest, baddest battle of them all, the war on cancer. And, given the toxicity of virtually all the drugs used to treat the Big C, you just have to think that in the end even Thomas Sydenham—the man who developed laudanum, but feared colchicum—would have approved this newest effect of our big toe on our lives.

In the beginning, when we were not yet first among primates, our feet were still hands and that toe was still a finger, special in its opposability, but a finger nonetheless. As it evolved, moving into line with the other four, our third and fourth hominin hands became feet. We gained a platform on which to stand, but along with it came a site for the pain of gout. Podagra, now defined by *Stedman's Medical Dictionary* as "Severe pain in the foot, especially that of gout in the great toe," has been with us since the Ebers papyrus. It was in the Greek pantheon in the person of its namesake goddess; in the pipes that leeched lead into Roman homes and human bodies; in the British, French, and American dance around the Revolution; in our discovery of the chemistry of proteins and purines; and finally in our modern drug laws and regulations, one more example—if one were needed—of how our special human foot has drawn its outline on the canvas of our lives.

## Notes

(1) Chaplin was serious about film and movement, but not about gout. He included comical goutry characters in two films, *His New Profession* (1912) and *The Cure* (1917).

(2) If you are hearing impaired, your newly grafted toe will also restore your ability to communicate fully with sign language, a system that connects you to the Celts. American sign language is generally believed to have been invented by Laurent Clerc and Thomas Hopkins Gallaudet, but the pair, who set up the first American school for the deaf in 1817, were certainly not the first to codify hand signals. For example, *ogham* is an ancient Celtic alphabet based on indentations and lines for vowels and lines for consonants that may still be seen on very old monuments in Ireland; positioning the fingers to imitate the written letters gave the Celts a secret language invaders could not understand.

(3) The list of the arthritides (singular: *arthritis* from the Greek word *arthron* meaning *joint*) includes, but is not limited to:

| | |
|---|---|
| Ankylosing spondylitis | Basal joint arthritis |
| Aseptic necrosis | Behçet disease |
| Avascular necrosis | Bursitis |

Carpo-metacarpal arthritis
Calcium pyrophosphate
    dihydrate deposition
    disease
Carpal tunnel syndrome
Celiac disease
Costochondritis
Crohn disease
Degenerative joint disease
Dermatomyositis
Discoid lupus erythematosus
Ehlers-Danlos syndrome
Eosinophilic fasciitis
Felty syndrome
Fibromyalgia
Fifth disease
Forestier disease
Fungal arthritis
Gaucher disease
Giant cell arteritis
Gonococcal arthritis
Henoch-Schönlein purpura
Infectious arthritis
Inflammatory bowel disease
Joint hypermobility
Juvenile arthritis
Kawasaki disease
Legg-Calve-Perthes disease
Lupus arthritis
Lyme disease
Pseudogout
Psoriatic arthritis

Raynaud phenomenon
Reiter syndrome
Restless legs syndrome
Rheumatic fever
Rheumatoid arthritis
Scleroderma
Septic arthritis
Sjögren syndrome
Somatotroph adenoma
Spinal stenosis
Temporomandibular joint
    disorder
Mixed or undifferentiated
    connective disease
Marfan syndrome
Mycotic arthritis
Osgood-Schlatter disease
Osteitis deformans
Osteoarthritis
Osteonecrosis
Osteoporosis
Page disease
Palindromic rheumatism
Polyarteritis nodosa
Polymyalgia rheumatica
Polymyositis
Temporal arteritis
Tietze syndrome
Tuberculous arthritis
Ulcerative colitis
Vasculitis
Viral arthritis

(4) Bocking and de Wyche served an authentic architectural treasure, the
Chichester Cathedral. The building was begun in 1078, consecrated in
1108, rebuilt in 1178, given a new Christopher Wren spire in 1721 to
replace one destroyed by a lightning strike, and rebuilt again in 1861 after
the original stone walls collapsed. Today the Cathedral is famous not
only for its history, but also for its art, a modern collection commissioned
by Walter Hussey (1909–1985), Dean at Chichester from 1955 to 1977.
Hussey was clearly a man of catholic taste, embracing arts and artists
of many persuasions. His collection includes Leonard Bernstein whose
*Chichester Psalms* (1965) had music from West Side Story and Hebrew
text from the Old Testament; Mark Chagall whose stained glass window
(1978) was based on Psalm 150 (". . . let everything that hath breath praise
the Lord"); John Piper, whose tapestry of the Holy Trinity was woven

in France in 1966; and neo-romantic British artist Graham Sutherland, known for his Welsh landscapes but represented here by an altarpiece, *Noli Me Tangere* (1961), with Christ holding out his hand to Mary Magdalene.

(5) To convert degrees Celsius to degrees Fahrenheit, multiply the Celsius number by 1.8 and add 32. To convert Fahrenehit to Celsius, subtract 32 from the Fahrenheit number and divide the result by 1.8.

(6) As William Safire explained in *The New York Times* in 1997, the sobriquet came from "white 'bucks,' the casual, carefully scuffed buckskin shoes with red rubber soles and heels worn by generations of college men at Ivy League schools. Many of these kids, supposedly never changing their beloved footgear, went on to become masters of the universe on Wall Street and in the best-known law firms."

# 5

# DESIRE

'They are not wise, then,
who stand forth to buffet against Love;
for Love rules the gods as he will,
and me."

Sophocles, *Trachiniae (The Women of Trachis, c.* 430 BCE)

IF THE *Vitruvian Man*'s is the ideal masculine form, with every part, private and otherwise, exquisitely detailed, then Cinderella's is the perfect female foot. Generations of women may have wondered how the girl managed to get around without cutting a toe on that glass slipper, but not a single one has ever doubted the sublime femininity of the foot that fit the shoe too small for the nasty stepsisters whose loathsome nature is captured in the size of their own unfortunate extremities.

The Aarne-Thompson-Uther (ATU) Index is a system that classifies traditional folk tales by plot. Created by Finnish folklorist Antti Aarne (1867–1925) in 1910, it was expanded and translated into English in 1928 by Stith Thompson (1885–1976), professor of English and folklore at Indiana University, and then updated in 2004 by Hans-Jorg Uther, professor of German literature at the University of Duisburg-Essen. In all three versions, *Cinderella* is listed as ATU 510 A, a subtype of the category ATU 501 (persecuted heroine), one section of *Supernatural Helpers 500–559*.

East, west, north, and south virtually every culture on earth has its own version of this Mistreated-Motherless-Girl-Finds-Prince story. Some estimates put the number of such tales at more than 1,000, some with shoes, some without, some with cinders, some without, but all with a motherless, orphaned, or abandoned heroine, an animal guardian, a magic gown, and a handsome prince.

One very early example is *Rhodopis* (*Rosy Cheek*), the fictionalized sixth century BCE tale of a real woman variously described as a Thracian courtesan or a slave, but either way a stranger and alone in Egypt. As recorded by the Roman historian Strabo, this girl gets her prince when a bird picks up one of her sandals, flies to the palace, and drops the shoe in the lap of the Pharaoh, Amasis, who immediately orders a search for its owner whom he soon finds and makes his queen.

The Chinese Cinderella girl, *Yah Shen/Yeh-Hsein*, has a magic fish carry her shoe to her prince. The no-shoes *Katie Woodencloak* (Norway) has a friendly bull; the heroine of *Ashey Pelt* (Ireland), a black ewe; in *Rashin-Coatie* (or *Rashiecoat*, a Scottish king's daughter made to dress in a *rashin coat*, a garment made of grasses), a red calf; in *Conkiajgharuna* (*The Little Rag Girl*, Georgia), a cow. *Nomi*, the Zula Cinderella, follows a talking fish's instructions to eat him and then toss his bones into the garden of the chief, who then seeks to choose a wife for his son by asking the women of the village to pick up the bones—which slip out of everyone's hands but Nomi's who, thanks to her finny friend, walks off with the prince.

After years of making do only with helpful animals or an occasional older woman, the Cinderella girl—in this case, the Parisian Cendrillon—finally got her fairy godmother in 1697 courtesy of Charles Perrault (1628–1703), a member of the Academie Française who also introduced the pumpkin that turned into a carriage attended by mice that turned into men . . . and the wonderful glass slipper. Some histories of the Cinderella shoe story suggest this last may have been an accident of language, that when Perrault first heard the tale he confused the word *vair* (French for squirrel fur) with *verre* (French for glass). Others dismiss this as a literary urban legend.

Perrault's typically frothy Parisian romance was followed by *Aschenputtel* (*Ash Girl*), the Grimms' typically grim German version. Jacob and Wilhelm's persecuted heroine plants a tree on her mother's grave. The tree attracts a white dove who grants all the girl's material wishes including an increasingly lovely wardrobe of gold and silver dresses and gold and silver shoes for an increasingly important series of royal balls. Eventually, Aschenputtel gets her man and her gorgeous wedding, but it doesn't end pleasantly. The girl and the prince have barely said, "I do," when the Grimms, ever true to their name, conjure up an entire avian air force—white pigeons, turtle-doves, and "all the birds beneath the sky"—that, like creatures from a Hitchcock movie, dive-bomb the stepsisters and peck out their eyes.

After that, the only thing left to do was to liberate Cinderella from her need to find a protective Prince, a task addressed by therapist Colette Dowling in *The Cinderella Complex*, a treatise urging women to forget the prince and take control of their own lives. The book, published in 1981 at the height of the feminist awakening, has been translated into twenty languages, but the emergence of Chick Lit as a major category suggests that not all that much has changed. Around the world, little girls still fight with their mothers and sisters, consider themselves strangers in their own homes, and dream of escaping into an ending where the Girl still gets her Prince and the Prince still gets the Right Girl.

But is it the girl that he wants? Or simply her foot?

## The heart wants what the nose knows

Like the palm of your hand, the sole of your foot is an erogenous zone, highly innervated and exquisitely sensitive to touch. It is also perfumed by friendly bacteria that digest and then excrete the produce of 250 thousand sweat glands per foot, bathing your feet in a salty sweat that carries messengers called *pheromones* (from the Greek words *pherein* meaning *to carry* and *hormon* meaning *to impel*).

Pheromones were isolated and identified in 1959 by Nobel laureate Adolf Butenandt (Chemistry; 1939) and named by German biochemist Peter Karlson and Swiss entomologist Martin Lüscher. These chemicals, first found in silkworms, come in two broad categories: releaser pheromones and primer pheromones. The releasers initiate a behavioral response in another individual, such as causing him or her to move closer to the source of the delicious odor. The primers set off a physical response, such as an increase in another's blood pressure or heartbeat or the secretion of sex hormones. Together, the two molecules may explain the mysterious phenomenon of love at first sight that some suggest might more accurately be labeled, "love at first scent"—the caveat being that you can *see* across a crowded room, but you have to be pretty much nose to nose to catch the scent.

Insects get their olfactory message, including the messages from pheromones, through their antennae; animals get theirs through olfactory receptors (OR) located on the vomeronasal organ (VNO), two structures located on the lining of the nose or the roof of the mouth and found in all animals, from dogs to snakes to us. The first stage of the accessory olfactory system, the VNO contains sensory neurons that detect chemical stimuli such as pheromones, which play roles in reproductive and social behaviors.

When we stood up on the dry African savannah or by the river as Alister Hardy insisted, our nose rose with us, our sense of sight replaced our sense of smell as our most important translator of information from the outside world, and like the rest of our body, our eyes adapted to our upright posture.

Our eyes are set into our skull so that they look forward, not off to the sides as a fish's or snake's eyes do. As a result, we have binocular vision, the ability to form a single image by merging two different images, one from each eye. To see how this works, look straight ahead at an object, say a pen, on the desk in front of you. Now, while looking at the pen, close one eye, then open it, and close the other. As you do, the pen will seem to move to the left when you close your right eye and to the right when you close your left eye. But when both eyes are open, the two images become one, creating the stereoscopic effect, a perception of depth that enables you to judge the size of the pen and how far away it is.

We are not the only ones with binocular stereoscopic vision. Many predators, who live in a world where the governing rule is "See, catch, eat," also have forward-facing eyes with binocular stereoscopic vision. On land, lions are a good example. Their prey—think antelope—are at a disadvantage, relying on shadows or images glimpsed through eyes positioned on the sides of the head to avoid being captured and consumed. In the sky above us, raptors such as eagles and hawks also have eyes that look forward, and they see even better than we do. These killers of the air can spot a rabbit or rat several miles on the ground below and then dive with absolute accuracy to pick it up; nocturnal raptors such as owls can do it in the dark.

Perhaps because of our evolutionary acute vision, our VNO became vestigial, meaning it was once useful, but no longer seems to be. This would not be the first such loss. The appendix that may once have played a role in our digestive system and the "wisdom teeth" that once nestled into a significantly larger hominin or early human jaw are also vestigial. The coccyx, a.k.a. our "tailbone," may also seem vestigial. It isn't because no human ever had a true tail. Early in pregnancy, all human embryos do have one, a reminder of our evolutionary development that recedes and disappears as our lower limbs develop. Very rarely, a baby is born with soft tissue that looks like a tail protruding from the base of his spine, but

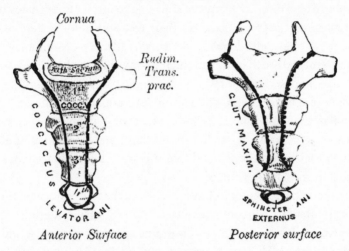

Anterior Surface                      Posterior surface

the structure, known as a *caudal appendage,* is either a failure of the lower limbs to develop properly or a neural tube birth defect that may occur alone or together with other problems such as a missing limb or cleft lip and/or palate. Like several other genetic defects, such as clubfoot, the caudal appendage may run in families, but the coccyx itself is a useful part of our anatomy, part of the boney frame on which we sit when we recline, as well as the site where muscles that support the pelvic floor and move the upper leg are anchored.

As for the VNO, other mammals such as dogs have an organ that may comprise as much as sixty square inches covered with tiny hairs called cilia and hold as many as 300 million olfactory receptors along with nerves that send fibers from the VNO receptor sites straight to the brain which decodes the pheromone's messages. By comparison, our human VNO covers only one square inch with as few as 5 million or so receptors, and we have no such nerve connectors. To paraphrase neuroscientist Michael Meredith of Florida State University in Tallahassee, the genes in the human VNO receptors are pretty much dead as dodos, defective pieces of DNA that look like genes, but do not transmit messages. In 2004, researchers at the Max Planck Institute for Evolutionary Anthropology in Germany and the Weizmann Institute in Israel labeled about 60 percent of the genes in the human VNO (and 30 percent of those in the nonhuman primate VNO) as pseudogenes.

However, even given our visual advantage and our dodo genes, it would be wrong to dismiss the influence of our less acute sense of smell and our sensitivity to pheromones on our lives and loves.

Whole groups of other primates continue to attract their mates by rubbing testosterone-scented urine on their feet and hindquarters. Animals such as female cats and dogs release a scent attractive to the males when they (the females) are in heat. Among humans, there is the still-classic phenomenon first described in 1971 by researcher Martha McClintock of the University of Chicago, who while an undergraduate at Wellesley College, noticed that the menstrual cycles of the women living together in her dormitory were synchronizing. In her senior thesis and a paper later published in *Nature*, McClintock concluded that this was due to pheromones transmitted from woman to woman.

More recently, cell biologist Claus Wedekind has demonstrated the influence of the major histocompatability complex (MHC), a group of genes whose primary job is to enable our immune system to recognize and rebuff foreign invaders. MHC also plays a role in creating body odor and setting our preference for one odor over another. In choosing their mates, mice prefer those with MHC different from their own. Several studies, including those run by Wedekind, appear to show the same effect in human beings.

Wedekind begins his studies by defining the MHC profiles of his male and female volunteers. Then the males are asked to live as "odor neutral" as possible—no cologne, no perfumed soaps, no perfumes detergents, no odorous foods, no alcohol, no smoking—for forty-eight hours and to wear the same T-shirt to bed for the intervening two nights. After that, the female volunteers are asked to describe the odors left on the T-shirts. Clearly, opposites attract—perhaps Nature's way of protecting the species from the problems experienced by the intermarried Egyptian pharaohs and European royals—because like the mice, the women were more attracted to the odors left by men whose MHC profile was different from their own. The scent, they said, often reminded them of current or past partners. There seems to be a hormonal basis for the choice: The responses were reversed when women were using oral contraceptives.

So who's to say that the odor of our smelly feet is not also a sexual signal?

Certainly not Sigmund Freud.

Say what you will about his apparent problem with women, the man was not squeamish.

Every dog and cat owner knows that animals find other animals' nether parts seriously interesting, probably as a means of identifying friend or foe. As University of British Columbia psychologist Stanley Coren writes in *How Dogs Think: Understanding the Canine Mind*, a dog's nose "not only dominates his face, it also dominates his brain [and] thus his picture of the world."

Freud declared that with our noses high above ground our natural animal "coprophilic instinctual components" became "incompatible with our aesthetic standards of culture." But for some, the brain does not accept the message. *Coprophilia,* from *kopros*, the Greek word for *feces*, and *philia* which means *love* or *attraction*, retains a secure place in the catalogue of human erotic behavior. Freud invented a word—*drekological* from *drek,* the German term for mud or excrement—to describe the practice that one standard medical school text, *Kaplan and Sadock's Concise Textbook of Clinical Psychiatry*, defines as "sexual pleasure associated with the desire to defecate on a partner, to be defecated on, or to eat feces (coprophagia)."

This is generally regarded as repellent, but human manners—Freud's "aesthetic standards of culture"—are no match for human instinct. You can deny a child the tactile and olfactory pleasure of finger painting with his feces, but as an adult he (or she) will still be attracted to the sexy, smelly parts of another person's body, sufficiently so, one hopes, to reliably insure another generation. As French psychiatrist Dominique Arnoux writes in *The International Dictionary of Psychoanalysis,* Freud concluded that "[r]epression of an olfactive coprophilic pleasure can determine the choice of a fetish." To be more specific, in February 1910, Freud wrote to his colleague, Berlin psychiatrist Karl Abraham (1877–1925), "I regard coprophilic olfactory pleasure as being the chief factor in most cases of foot and shoe fetishism."

## Do gods have feet?

The Egyptians, Greeks, and Romans were polytheists, on friendly, some-times intimate, terms with their gods. For example, Hercules is a demigod, the son of Zeus and Alcmene, the wife of Amphitryon, king of Thebes, conceived on an evening when the king was away on business. Zeus visited the sleeping queen and then magically extended the night long enough to allow Amphytrion to return and climb into bed with his wife who nine months later gave birth to two boys, one (Hercules) immortal and the other (Ipicles) not. Thanks to Zeus' considerate time-shifting, the queen remained unaware and therefore not adulterous, thus avoiding a major royal mess in Thebes.

A few of the early Mediterranean gods had unusual body parts, such as Pan's goat legs and horns and the jackal head sported by Anubis, the Egyp-tian god of the dead. But as a rule, while the Egyptian's deities were taller and slimmer and the Greeks and Romans more muscled and voluptuous, they looked like you and me, or in the case of the Greeks and Romans like what you and I would look like if we had been sculpted by Praxiteles.

The introduction of monotheism changed the picture by eliminating it. There was now only one God: no more major and minor deities. There was also a new prohibition against graven images, and most important, the God of the Old Testament was invisible. "No one," He tells Moses on the mountain, "can see me and live."* True, his shape was assumed to be human because, as it says right there in Genesis 1:27, "God created man in his own image." And, yes, the author of Genesis must have meant "man" to signify "mankind," that is, man *and* woman, because in Hebrew there are several names for God, some male, some female, and some plural, which, as you can imagine, would have further complicated the image issue. Even today, observant Jews do not draw pictures of their God, a prohibition also embraced by Islam. The only representations are symbolic such as the engraving of a hand sometimes found on a Jewish gravestone.[1]

Christian monotheism re-introduced the pictures and statues, perhaps in an attempt to reconcile with and welcome polytheistic pagans via a familiar image. Leaf through the portraits of the Christian God, and you find that they all tend to look like the people living where the artist lived. Early on, that meant young, middle-age, or older Eastern men. As time

---

\* Except where noted, the biblical phrases used here are from *The Holy Scriptures,* The Jewish Publication Society of America, Philadelphia, 1712–1952 or The Holy Bible, King James Version, American Bible Society, New York, 2009.

passed and Christianity spread, the pictures began to look like Europeans, with skin lightened to please a Western sensibility, or at least a Western European one, leading inevitably to the obvious question from various converts: "When did Jesus become white?"

The Bible does not leave us completely without guidelines. We do know that its God is a "dwelling place" underneath which "the everlasting arms" of Deuteronomy 33:27, stretch wide enough to encircle the world. We know that at the end of the arm there is a really big hand because, as the American spiritual assures us, "He's got the whole world in his hands," and that the hand of God, as celebrated by Michelangelo in the Sistine Chapel, confers life and intelligence on man.

We know, in short, that this God's hands are open, welcoming, loving. But here's the important part: His feet are not only invisible, they (and ours) are symbols of far more aggressive traits.

Among the ancients, feet figure prominently in accounts of one deity's, person's, or nation's establishing its authority over another. We never "see" the feet of God in the Bible, but we are frequently told what's *under* them: The beautiful sapphire pavement (sky) Moses and the elders of Israel saw when they went to collect the Ten Commandments (Exodus 24:10), the not-so-beautiful storm clouds "like dust under God's feet" (Nahum 1:3), and the definitely subservient earth as God's footstool (Isaiah 66:1). To stamp man as his emissary, the monotheist God puts "all things under his [man's] feet" (Psalm 8:6). In First Kings 5:3, King David cannot "build a house unto the name of the Lord his God for the wars which were about him on every side, until the Lord put them [the enemy] under the soles of his feet."

But if the authors of the various books of the Bible were shy about describing God's feet, they are positively Chatty Cathy when it comes to ours, which, Old Testament and New, are obviously more than a simply 10 toes, 52 bones and 66 muscles on which to stand.

Early on, people rarely wore shoes indoors, especially indoors at a place of worship. Stepping forward to address the congregation, Jews and Christians went barefoot in accordance with the dictates of not one, but three, books of the Bible, Exodus 3:5s, Joshua 5:15, and Acts 7:33: "Put off thy shoes from off thy feet, for the place whereon thou standest is holy ground." This gesture of respect is also honored in Islam, among Hindus, and by Sikhs at their houses of worship, respectively, the mosque (from the Arabic *masjid* meaning *place of worship*), the *mandir* (from the Sanskrit *mandira* meaning *dwelling*, i.e., of the deity) , and the *gurdwara* (Pakistani for "gateway through which the guru can be reached").

As for face-to-face human relationships, Biblical scholar Richard D. Patterson notes that records from ancient Mesopotamia show us that "the Neo-Assyrian king Ashurbanipal frequently speaks of the submission of his enemies as 'kissing his feet.' For example, he boasts that the Elamite king Tammaritu 'kissed my royal feet and smoothed (brushed) the ground (before me) with his beard.' A similar idea occurs in the texts of ancient Egypt. Thus, the victory hymn supposedly coming from the god Amon Re to Thutmose III declares, 'I have felled the enemies beneath thy sandals.' "

Biblical soldiers are said to have followed the example of these earlier warriors, drawing images of their enemies on the soles of their sandals so as to grind them into the ground with every step. Losers were to be made footstools for the victors (Psalm 110:1) or forced to bow "with their face toward the earth, and lick the dust" off the winners' feet (Isaiah 49:23).

After which, the winners would likely wash their feet and maybe their sandals too. Sooner or later that could easily become a ritual: Stomp the enemy, wash your feet.

Certainly, foot washing (*pedilavium* in Latin) was more than a victory dance in the end zone. If you live in a desert, rinsing your feet before entering someone's home is a common courtesy to keep from tracking sand and mud all over the floor or carpet. In *Genesis,* considerate hosts like Abraham and Lot provide water for the guest and even a servant to do the bathing. Once inside, a visitor was careful not to disrespect his host by showing the bottom of his foot, clean or otherwise. In many Asian and Middle Eastern countries, the shoe is also unclean. Tossing one at someone remains the ultimate insult as an Iraqi television correspondent made clear during a goodbye press conference in Baghdad on December 14, 2008, when he hurled not one, but two, shoes at then-President George W. Bush, accompanied by the eloquent curse: "This is a farewell kiss, you dog!"

The Old Testament presents washing a visitor's feet as a clear sign of respect. In First Samuel 25:41, King David's soon-to-be second wife, Abigail, offers to wash the feet of her soon-to-be-second-and-seriously-influential husband's servants. In The New Testament, the practice became a symbol of equality when, after washing his disciples' feet, Jesus told them to do the same for each other "to prove that a servant is not greater than his master; nor is he who is sent greater than he who sent him," as clear a statement of human rights as the one 1,743 years later that begins, "We hold these truths to be self evident . . ."

This was all such good public relations that the Church adopted the alms and bathing idea, christening it "Maundy" from the Latin word *mandatum* meaning *command*, as in Jesus' command to his disciples. And what the Pope did, the King of then-Catholic England could not ignore. In 1211, King John I took a pass on the bathing part, but handed out alms in an attempt to placate Britons angered by his having raised taxes to ransom his brother Richard the Lionheart who had got himself kidnapped at the end of the Third Crusade—and also to make nice to Pope Innocent III who had excommunicated the king for seizing church property. Edward I, who ruled from 1271–1307, formally designated the day before Good Friday as Maundy Thursday. Both Roman Catholic Mary Stuart and her sister, Church of England Elizabeth I, also washed the feet of some deserving poor, after said feet had been thoroughly rinsed by Court officials. The last British ruler actually to splash water on impoverished feet was James II in 1689. The task of handing out alms was handled by Court flunkies until 1931, when George V decided to do it himself, awarding "Maundy money" each year to as many deserving men and women as he was old. His granddaughter, Elizabeth II, has continued the tradition; in 2012, age 86, she distributed two purses each to 86 men and 86 women, one red purse with coins representing such essentials as food and clothing and one white purse stuffed with commemorative coins.

The British Royals may have abandoned bathing naked feet, but the Pope still performs the ritual in Rome, and in appropriate situations so do other Church officials seeking to demonstrate repentance and humility. In February 2012, at Saint Mary's Pro-Cathedral in Dublin, Dublin Archbishop Diarmuid Martin and Cardinal Sean O'Malley of Boston knelt to wash and dry the feet of eight men and women who as children had been victims of clerical abuse.

Which brings us in a very roundabout manner to the moment—the act—where we all began.

As anyone who has ever read the book can testify, the Bible is awash in the erotic. Not just the lush and liquid Song of Songs; the whole thing, cover to cover, with so many references to your feet as symbols that you can practically hear the chorus in the background muttering, "Dirty, smelly, hidden—sex!"

Before Eden, shame was not unknown, but as you might expect in a warm environment where people ran about scantily clad, shame about one's body was uncommon. It was only after Eve that the naked human body became a sign of lost innocence, to be covered with "garments of skin" (Genesis 3:17) or even with something as minimal as an artistic fig leaf. Ditto for the naked human foot. Being the body part closest to the ground, it was both good (your connection to the earth that provides sustenance) and evil (your connection to the cursed earth), and thus a useful stand-in for body parts and functions considered unmentionable in polite society.

Neuroscience and the PET scan have made it possible to show how the areas of our brain light up when we perform specific intellectual tasks such as solving math problems (women appear to use both sides of the brain; men, only one) or experience highly emotional moments such as thinking about a loved one, but in language and mythology, we still attribute character traits to particular body organs, especially the heart. We may scoff at those ancients who ate the heart of a conquered enemy to acquire his courage, but we are equally if less bloodily entranced: We say that a successful athlete has heart, and every St. Valentine's day we glorify the heart as the place where love lives.

*People of the Body* (1992) is a collection of essays subtitled, *Jews and Judaism from an Embodied Perspective*. In the essay, "Images of God's Feet," New York University professor of Hebrew and Judaic Studies Elliot Wolfson describes the use of feet as biblical symbols for the male genitalia, explaining that repeated references to putting on and taking off sandals, bathing one's feet, and uncovering the feet are implicit references to sexual intercourse as in the Book of Ruth 3:4-6 when Naomi, instructs her widowed daughter-in-law Ruth to visit a sleeping Boaz: ". . . when he lieth down, though shalt mark the place where he shall lie, and thou shalt go in, and uncover his feet, and lay thee down." There is also the ancient Jewish ceremony of Haliztah ("taking off") described in Deuteronomy 25:7-10. The ceremony relieves the brother of a man who has died childless of his obligation to marry the widow and releases the widow to marry someone else. The two come before the rabbinical court (*beth din*) where she says he refuses to marry her, he confirms it, she removes his right shoe, throws it down, spits on the ground in front of him, and everyone in the room knows exactly what that means because they know that, like uncovering the feet, putting a foot into a shoe is shorthand for sexual intercourse and her removing his shoe says this did not/will not happen.

Just as *uncovering the feet* is a euphemism for sexual intercourse, *covering the feet* is a euphemism for urination, as in Judges 3:24 when servants, knocking on Ehud's locked bedroom door, tell each other that, "Surely he is covering his feet in the cabinet of the cool chamber." And for anyone who missed the meaning, the evolution of the biblical language from classic to modern makes it perfectly clear. The phrase "Saul went in to cover his feet" (First Samuel 24:4) in the original Old Testament and the King James Bible is "Saul went in to relieve himself" in the *New International Version (NIV)*.

Using feet as symbols for genitalia and bodily functions actually makes metaphorical sense. Our bipedalism not only facilitated our sexual display, but also allowed us to engage in sex standing up, face to face, a position that requires serious athletic ability and is so fraught with conflicting emotions that authors such as Mario Puzo in *The Godfather* and filmmakers such as Bernardo Bertolucci in *Last Tango in Paris* have embraced it as a visual image for one or both of the extreme poles of sexual encounter—urgent passion and uncontrollable violence.

And, yes, you can certainly carry this kind of analysis too far.

It is, unlikely, for example, that feet are metaphors for female genitals. Some see that possibility in passages such as Deuteronomy 28:57 when during childbirth, the child "cometh out from between her feet" or Ezekiel

16:25, a long denunciation of Jerusalem, the wayward Bride of God: "Thou hast built thy lofty place at every head of the way, and hast made thy beauty an abomination and hast opened thy feet to every one that passed by, and multiplied thy harlotries." But the first seems a simple description of one position during labor and the second less a substitute for lady parts than a plain accusation of wifely betrayal: "I gave you everything you asked for and you still chose to go clubbing and flirt with other men instead of paying attention to me." The choice of language in succeeding versions of Ezekiel appears to prove this true. In the Jewish Publication Society of America Old Testament (1952), there are nine "harlots" and seven "harlotry/harlot-ries." In the King James (1611), the "harlots" are still there, but the '"harlot-ry/harlotries" have become the less poetic "fornication" and "whoredom." By the time the NIV was first issued in 1971, there are no harlots and no harlotry. What you see is what you get: "Prostitution."

## Uncovering what's inside

In *The Book of Filial Piety*, Confucius (551 to 479 BCE) says that the human body is precious because it is a gift from one's parents: "Our bodies, to every hair and bit of skin, are received by us from our parents, and we must not presume to injure or wound them."

Others in the ancient world—which for the purposes of this chapter means the Mediterranean cultures in the centuries before the fall of the Roman Empire—believed that the body was a receptacle for the spirit of its god(s). This notion certainly predates monotheism, but the actual words are perhaps written most definitively in *First Corinthians 6:19-20*: "Know ye not that your body is the temple of the Holy Ghost which is in you?"

Add to this a belief in life after death, and you can see that keeping the dead body intact becomes really important in some circles. Chopping off parts of the losers' bodies after battle was a fairly universal custom, but the corpses of one's own tribe, civilians and soldiers alike, were treated with respect. As University of Manchester's Jacqueline Finch explained in her description of the Egyptian artificial toes, one of which was found on a mummy, Egyptian "embalmers made every attempt to reinstate the com-pleteness of the physical body before burial. The body could be moulded in plaster, packed with mud, sand, linen, butter, or soda. Even sawdust was stuffed between the skin and muscle to reform the contours, and false eyes, noses, and often genitals were added. Where limbs were missing generally poor imitations were added, of linen, reed, mud, and resin. Spells inscribed on the walls of ancient Egyptian royal tombs (known as the Pyramid Texts,

c. 2375 BCE), spells decorating the inside of coffins (the Coffin Texts, c. 2055 BCE), and mortuary texts written on papyrus (the Book of the Dead, c. 1550 BCE) all refer to the importance of 'reassembling' and 'reuniting' the body to enable 'revitalisation' to take place in the Afterlife. This was seen as a prerequisite to the mystical reanimation expected in the next world."

The Jews forbade cremation, insisting that a body must be buried whole. Christians thought resurrection might be impossible if parts were missing. Even today, in some cultures some people still save "missing" tissues such as an excised appendix or gallstone or a bloodied cloth in special containers for eventual burial with the rest of the body. Although Islam and Judaism remain formally opposed to autopsy, organ donation is now most commonly accepted by all religions as an act of human kindness in keeping with their teachings so long as it does not hasten the death of the donor— a touchy question for those who believe that death arrives not when the brain ceases its activity, but when the heart stops beating.

Human dissection has always been—pun intended—an even more dicey subject, but here it leads by twisty path from the visible feet of the Mediterranean gods to the invisible feet of the God of the *Bible* and then to the Christian God of Michelangelo's Sistine Chapel whose feet are plainly visible, but whose brain is hidden in plain sight.

In medicine and art, how we see our bodies depends to a large extent on how *clearly* we see our bodies, including the feet on which we stand. What we once imagined to be hidden under the skin, based on animal dissection or pure fantasy, could be displaced only when we began to cut into the human body itself, and doing that required us to reconcile medical necessity with our spiritual and intellectual biases, a sometimes uncomfortable process.

As Sushruta, the seventh or sixth century BCE Indian surgeon, wrote in the *Sushruta Samhita* (*Sushruta's Summary*): "Any one, who wishes to acquire a thorough knowledge of anatomy, must prepare a dead body and carefully observe and examine all its parts." Forbidden by his Hindu faith to cut into the body with a knife, Sushruta came up with an acceptable alternative, submerging a corpse in water for a week or more to soften and decompose the tissues which soon fell away to reveal the structures underneath.

The first medical dissection of a human body, complete with scalpel, appears to have been performed in the late sixth century BCE at the medical school in Croton, a city-state in Magna Graecia (Latin for *Great Greece*) at the bottom of the Italian "boot" where the people were known as Italians to the Greeks and Greeks to the Romans. Croton seems to have been far enough off the beaten philosophical path to permit a physiologist named Alcmaeon to slice away in peace, in the process identifying the optic nerve, differentiating veins from arteries, and naming the brain as the center of the intellect. After that, for centuries the internal anatomy of the human body remained, to steal a phrase from Donald Rumsfeld, the former U.S. Secretary of Defense, a "known unknown," as the acceptance of human dissection swung back and forth like a pendulum depending on where you lived and who was in charge. Aristotle is said to have performed two dissections in secret, but his *Parts of Animals,* written around 350 BCE, is based strictly on species other than us. Fifty years later, Ptolemy I (c. 360–c. 282 BCE)— former Greek general; Alexander's biographer; self-appointed Egyptian King; and founder of the Ptolemaic dynasty, which produced fifteen kings named Ptolemy who ruled Egypt for more than 300 years—encouraged two Greek physicians, Herophilus (335–280 BCE) and Erasistratus (304–250 BCE), to dissect human cadavers, and sometimes vivisect living human beings, at the school of anatomy they founded in Alexandria.

Much later, Claudius Galen (c. 129–c. 216), yet another Greek physician who went to practice in Rome, eventually rising to become court physician to the emperor, was rumored to have performed human dissections in secret. But Galen's anatomical drawings were based strictly on animal dissections, so they were often flawed. For example, his pictures of the attachment of the muscles in the upper back to the neck and head are simply wrong. As Raymond Dart knew when he first saw the Taung Child, a body that holds its head straight up (human), requires an arrangement of the supporting tissues different from a body whose head leans forward (ape and Neanderthal), or sits on a horizontal line with the spine (dog, cats, and other four footed creatures). Denied human dissection, Galen never saw the difference.

Christians formalized their opposition to dissection at the Council of Tours in 1163, but Renaissance artists—not Renaissance physicians—resurrected the study of anatomy. Art, not medicine, led the way to our true understanding of our anatomy, and art, not medicine, truly drew the picture of our foot, including the foot of God the Bible had ignored.

Although his anatomical drawings were not published until after his death, Leonardo ignored the Church and dissected as many as 30 human corpses. Clearly, Michelangelo also performed dissections. Nobody knows exactly how many, but evidently the bodies were mostly those provided by the Hospital of Santo Spirito in Florence in exchange for a wooden cross for the sacristy of the Church of Santa Maria del Santo Spirito, suggesting that while the religion decried dissection, many in the Church itself might not.

Of course, just as Galen's pictures of human anatomy had been flawed by his being limited to animal dissection, some of Da Vinci's and Michelangelo's representations of internal female anatomy were wrong because there just weren't enough female corpses available to make it possible for them to get the details right. That small issue aside, the artists' paintings, drawings, and sculpture, including the *Vitruvian Man*, are clear proof of how we benefited from their apostasy.

Dissection wasn't the only thumb they stuck in the Church's eye.

In Egypt, the gods were at least partially clothed. In Greece, male deities were often shown naked, but goddesses—except for Aphrodite—were clothed, although their garments left little to the imagination. The Romans showed both male and female deities unclothed; most famously Diana (the Roman name for Artemis) had one breast bared in imitation of Amazon warriors said to slice away one breast in order to draw the bow smoothly across the chest.

For Greek and Roman gods, "nude" didn't simply mean without clothes: it meant without imperfection like the *Vitruvian Man*. Not until the post-Classical Hellenistic period did Roman artists sometimes add wrinkles and other realistic signs of age to their human portraits such as *The Old Market Woman*, who stands (or stoops) forever ancient, one breast bare, in a first floor gallery at the Metropolitan Museum of Art in New York.

When the Reformation arrived on a wave of real and imagined scandal among the Catholic hierarchy, naked bodies in Church art became a target, with the Sistine panels squarely in the bull's eye. In 1563, the Council of Trent attempted to bat away the Protestant charges of corruption by countering with new-found Catholic prudery, decreeing that henceforth "[e]very superstition shall be removed [from art, particularly Church art] ... all lasciviousness be avoided; in such wise that figures shall not be painted or adorned with a beauty exciting to lust ... there be nothing seen that is disorderly, or that is unbecomingly or confusedly arranged, nothing that is profane, nothing indecorous, seeing that holiness becometh the house of God."

Nobody wanted to tangle with a living Michelangelo, who was known to retaliate by painting an adversary's face onto the body of a devilish creature. But once he was safely dead and buried, Giovanni Pietro Caraffa (Pope Paul IV), who also authorized the publication of the infamous list of banned books known as "Index Librorum Prohibitorum," set up what has come to be known as the "Fig Leaf Campaign" due to the Vatican's hiring an artist to cover the Sistine genitalia with fig leaves that remained in place until late in the twentieth century when the entire work was cleaned and restored.

As for dissection, there were still a few bumps on the road to serious study.

In 1543, thirty-one years after Michelangelo climbed down from his scaffold, Andreas Vesalius (1514–1564) published his anatomical master-piece, *De Corporis Humani Fabrica (The Structure of the Human Body)*. Twenty-one years after that, the Inquisition, which had obviously been busy with other things, sentenced him to death for ignoring Church dicta. But Vesalius, court physician to the Holy Roman Emperor Charles the Fifth and the Spanish King Philip the Second, had friends in high places. The death sentence was lifted, and he was sent off on a pilgrimage to Jerusalem instead. Alas, that turned out to be his own appointment in Samarra. He died when the ship on which he was returning home sank off the coast of Greece in October 1564, one year before Britain's Prot-estant Queen Elizabeth I gave London's Royal College of Physicians the right to begin dissecting human cadavers.

Elizabeth's gift to medical anatomists was limited to the corpses of four criminals a year who had been hanged in London or within 16 miles of the city. In 1752, the British expanded the rule with what is commonly known as The Murder Act, a law legalizing medical schools' dissection of any executed murderer for the study of human anatomy. The practice had already made its way to the American Colonies, where, as Temple University/Beasley School of Law professor Harwell Wells notes, the state of Massachusetts had ruled in 1784 that there were two ways to dispose of the corpse of a person killed while dueling. You could bury it with a stake driven through the body, or you could hand it over to the anatomist for dissection, probably at the recently opened school of medicine at Har-vard. New York passed a similar law, minus the stake-through-the-body rule, in 1789; one year later federal judges were given the right to add post-mortem dissection to a criminal death sentence.

Two hundred and one years later, in 1990, with dissection long since an established part of every physician's training, a gynecologist in Ander-son, Indiana, named Frank Lynn Meshberger opened a book about Michelangelo to the page showing the panel in the Sistine Chapel known as *The Creation of Adam* and blinked, maybe more than once you might

imagine. What he saw there, hidden in plain sight around the head of God, was something no one else had seen in the nearly 500 years since Michelangelo painted the chapel ceiling: A perfect, anatomically correct image of the human brain, brainstem, and spinal cord with God's hand reaching through the prefrontal cortex, the highly developed part of the brain just under the forehead where thought lives. "People have said that what is being passed from God to man in the painting is the spark of life," Meshberger told *The New York Times* during an interview the following October on the day an article he wrote about the painting appeared in the *Journal of the American Medical Association*. "But Adam is already alive. I think what God is giving to Adam here is intellect."

This would certainly fit right in with Michelangelo's views of an artistic inspiration as recorded in one of his own sonnets, which Meshberger included in his article:

> After the divine part has well conceived
> Man's face and gesture, soon both mind and hand,
> With a cheap model, first, at their command,
> Give life to stone, but this is not achieved
> By skill. In painting, too, this is perceived:
> Only after the intellect has planned
> The best and highest, can the ready hand
> Take up the brush and try all things received.

Others—art experts as well as physicians—were skeptical. You could hardly blame them. What makes the arts so interesting is that enjoying them is an intensely personal experience. Everyone agrees that the first four notes of Beethoven's *Fifth Symphony* (*Symphony in C Minor*) are three short *g*s and one long *e*: Dah-dah-dah-daaaaah. But how long? How short? What you hear depends not only on how fast or slow, loud or soft, the orchestra plays, but also on the condition of your ears, young or old, sensitive or not, clean or blocked by an allergy or head cold. Similarly, every painting is a Rorschach of its own, interpreted by what you feel and think as well as what you see.

Having published what *he* saw, Meshberger, like Raymond Dart after finding the Taung Child and Gregor Mendel after delivering his report on his garden, waited for someone else to look at the Michelangelo panels and say, "I see what he saw." He was lucky. It had taken Dart and Mendel half a century each to win the approval of their peers. Meshberger made it in twenty. In May 2010, after two decades of back and forth, yes and no, among art historians and doctors, Johns Hopkins neuroscientists Ian Suk and Rafael Tamargo published a piece in the scientific journal *Neurosurgery* saying they, too, had seen an image of the brain, brainstem, and spinal cord running up God's chest and throat in another Sistine panel, *The Separation of Light from Darkness.*

So, to summarize, history and the various books of the Bible tell us that the human foot (and its antecedent, the invisible foot of God) is a symbol of authority, dominance, subservience, humility, beauty, and bodily functions.

Who would not relish such a body part?

## Objects of desire

Unlike our pets, who love us for our inner beauty and the fact that we feed them every day, human beings often have preferences in body parts that dictate who we find attractive and therefore lovable. Women tend to focus on the whole picture ("He's tall" rather than "He's got long legs"). Men are more likely to aim for specific areas like breasts or legs or hair, the last a favorite of poets like Oscar Wilde (*Requiescat, Serenade, In the Gold Room*), who seems particularly taken with the golden variety, as was Dante Gabriel Rosetti describing Adam's first wife, Lilith, "whose enchanted hair was the first gold" (*Body's Beauty*), and James Joyce with his simple, "Lean out of the window, Goldenhair" (*Golden Hair*). Black is also beautiful: Byron celebrates the "raven tresses" of that woman who "walks in beauty like the night"; Charles Baudelaire, the *"Cheveux bleus, pavillon de ténèbres tendues"* ("Blue-black hair, a pavilion hung with shadows") in *Fleurs du Mal,* and Alfred Noyes, the long black hair of Bess, "the landlord's black-eyed daughter" in his romantic poem "The Highwayman."

No wonder some cultures insist that a woman cover her head.

There is no such direct command in the Old Testament, although Isaiah 47:2 does say that the "virgin daughter of Babylon" (that is, the Babylonian empire in the persona of a queen) will be shamed by uncovering her hair. The New Testament seems to offer conflicting advice. First Corinthians 11:5-6 says that "every woman that prayeth or prophesieth with her head uncovered dishonoreth her head" (two verses earlier, her husband is designated as *her* head, Christ as the head of every man, and God as the head of Christ), yet a few lines later at First Corinthians 11:15, we are told that a woman's long hair is itself a "covering."

Hair covered or not, some gentlemen do prefer blondes, and some women do opt, like Edna St. Vincent Millay, for the "brown hair that grows around [his] brow and cheek," but generally we love whom we love, sometimes for and sometimes despite their physical appearance.

Unless something interferes.

A fetish (from the Portuguese *feitiço* meaning *charm* or *magic*) is an object imbued with magical power like a rabbit's foot that conveys luck or a religious medal that protects against evil. French psychologist Alfred Binet (1857–1911) was the first person to use the terms *sexual-* and *erotic fetishism* to describe arousal triggered by a specific body part rather than by a whole person, or by ritualized behavior or costumes such as those gently pornographic scenarios featuring girls dressed in nurses' uniforms playing a grownup version of *Doctor.* The problem, if it is a problem, appears to affect many more men than women, although John Bancroft, Director of The Kinsey Institute from 1995 to 2004, a member of the International Academy of Sex Research, and author of *Human Sexuality and Its Problems*, suggests that the gender difference may be due at least in part simply to the fact that it is easier to identify male reactions to sexual stimuli via the visible erection.

Either way, in 2007, when five researchers from Department of Psychology, University of Bologna, the Group for Interdisciplinary Cultural Research at Stockholm University, the Zoology Institution in Stockholm, and the Department of Experimental Medicine at L'Aquila University (Italy) pooled their resources to see what sexual fetishes are most popular and with whom, you could say they were looking for an answer to the question Freud never asked: "What does a *man* want?"

They began by compiling a list of potential volunteer subjects, in this case by collecting a list of 2,938 English language online *Yahoo!* discussion groups whose name or description included the word *fetish*. After eliminating groups with non-sexual fetishes such as rock bands, they had 381 sites that fit their basic criterion: People "clearly identifiable as discussing a sexual topic." As for the inevitable questions about the reliability of a project using the Internet as a source when seeking volunteers, they concluded that "[s]ampling biases in Internet studies are often attributed to the higher socio-economical and educational status of Internet users. These, however, are no longer an elite in many countries, and it is estimated that 60% of USA citizens are Internet users. Although it is difficult to ascertain whether the putative 1,500,000 *Yahoo!* groups subscribers here surveyed represent the general population, it should be acknowledged that most of the research on atypical sexual behavior is based on data sources that are, in all likelihood, even less representative."

With a final list in hand, the Bologna researchers asked people to pick their pleasure, that is, to choose from the following list the body parts and features they found sexually arousing: feet and toes, body fluids, body size, hair, muscles, decorations such as tattoos, genitals, belly/ navel, ethnicity, breasts, legs and buttocks, mouth and lips and teeth, body hair, finger- and toenails, nose, ears, neck, and a seriously *drekological* candidate, body odor.

The runaway winner, no prompting required, was feet and toes which racked up 44,722 votes, more than five times as many as the next most popular item, body fluids, a category that includes blood, testimony to the sexual power and popularity of vampire tales from Bram Stoker's *Dracula* (1897) to Stephenie Meyer's *Twilight* series (2005).[2] Feet and toes were five times as popular as body size, which came in third; nearly seven times as popular as the hair on your head, which came in fourth; and eight times as popular as muscles, a sort of subdivision of body size, which was in fifth place. The surprises were the relatively unexciting genitals at number seven and breasts at number ten, followed by legs and buttocks at number eleven. Body odor was dead last at number eighteen, either because very few people wanted to admit an attraction or because it may be one of the things that makes feet important or because, as noted earlier, the genes in the human VNO receptors really are pretty much useless when it comes to recognizing pheromones.

As expected, most of the respondents in the Bologna study were male, but not as expected, most of them also chose feet and toes. The psychiatric rationale for this is classic Freud: Feet and toes resemble the penis, thus capturing male attention based on the universal and never-forgotten childhood fear of castration. But once again, you can go too far with such analysis—unless the place to which you plan to go is not the human psyche, but the human brain.

Vilayanur Ramachandran is the Director of the Center for Brain and Cognition and Distinguished Professor with the Psychology Department and Neurosciences Program at the University of California, San Diego, and Adjunct Professor of Biology at the Salk Institute. Where Freud saw penises in the relationship between sex and the human foot, Ramachandran sees synapses, connections in the body's primary locus of erotic response, the brain.

The Sensory Homunculus[3] is a drawing in which the size of various parts of your body are shown larger or smaller depending on the relative distribution of sensory nerves. For example, the well-innervated lips and tongue are shown much larger than some other supposedly more sensitive organs, rather like those maps on which New York is twice the size of the rest of the country. The sensory drawing was created by neurosurgeon Wilder Penfield (1891–1976). Like Fortunio Liceti, Penfield was an authentic over-achiever. Born in Spokane, Washington, he majored in literature at Princeton, won a Rhodes Scholarship to Oxford, graduated from Johns Hopkins Medical School, became a surgeon at Presbyterian Hospital (now New York Presbyterian Hospital-Columbia Presbyterian Center) and the associated Neurological Institute of New York. He then emigrated to Canada in 1928 to join the faculty of McGill University in Montreal and sign on as a neurosurgeon at what is now the Royal Victoria Hospital and the Montreal General Hospital at the McGill University Health Centre. Six years later, armed with a grant from the Rockefeller Foundation and the support of the governments of Quebec and Montreal, Penfield founded the Montreal Neurological Institute.

Penfield's specialty was surgery to relieve epileptic seizures. His approach was revolutionary. Rather than sedating the patient, he numbed the skull with a local anesthetic, then cut through and lifted off a piece of bone to expose the tissue underneath. Because the brain itself does not feel pain, Penfield was able to probe gently with the patient fully conscious so that his body responded in real time, enabling the surgeon to identify the exact location of seizure activity and then attempt to destroy or remove only that area. But Penfield didn't stop there. He began to map the sensory (and later the motor) areas of the brain, probing carefully and at each step asking the conscious patient what he felt or watching what he moved so as to establish the connection between specific areas of the brain and specific body parts. In 1951, he and another emigrant from the American Northwest, Herbert Henri Jasper (1906–1999), a psychologist, physiologist, anatomist, chemist, and neurologist, published their map in *Epilepsy and the Functional Anatomy of the Human Brain.*

What makes the Penfield/Jasper diagram fascinating—aside from the intelligent curiosity that led to its creation—is where in the brain you find some of the sites that control your various body parts. For example, the controls for the teeth, gums, and jaws are located right next to those for the lips. No surprise there. But the controls for the genitals, which is to say the *male* genitals—Penfield did not map the vagina—are next to the foot, not the upper leg. *Big* surprise. But an interesting one in light of (a) the connection between bipedalism and face-to-face sexual intercourse, and (b) Michelangelo's Sistine Chapel portrait of God with a surprise at the top of the body and another at the bottom, visible feet at one end and (for those who knew where to look) a visible brain at the other.

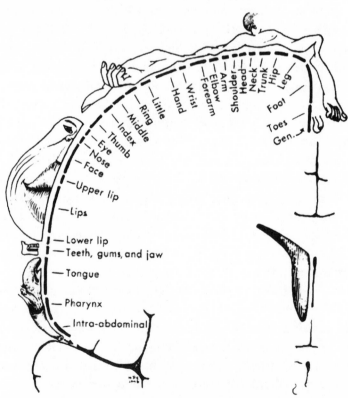

In California in the 1990s, Ramachandran was treating phantom limb syndrome, the phenomenon in which a patient feels pain or itching or simply the presence of a limb no longer attached to his body. Now, forty years after Penfield drew his sensory map by probing an exposed living brain, Ramachandran had a new, noninvasive search tool at his command to tell him why his patients felt what they felt.

The PET scan is an imaging technique that captures the activity of subatomic particles called positions emitted when your body consumes glucose, the fuel on which you run. The procedure begins with an injection of a mildly radioactive glucose solution that takes about an hour to make its way through your body and to your brain. When the scan is in progress, the image shows brightly lit spots at the sites where your brain is using glucose and releasing positrons. PET images are exceedingly subtle and complex. For example, your organs of hearing, speech, and vision, as well as the neurons associated with thinking, are each linked to a particular area in your brain. Technicians watching a PET scan following a real human brain reacting in real time to the physical act of eating and digesting food can also identify the areas of your brain that light up when you hear, see, say, or even *think* about the word *food*.

Neuroscientists use the term *plasticity* to describe the phenomenon by which sites in your brain adapt to the loss of some neurons or connections by switching their functions to other sites, a situation that can lead to unusual and unexpected couplings, which it turned out, were what Ramachandran's patients were experiencing.

As Ramachandran explained during an interview in 2011 on the National Public Radio program *Fresh Air*, when a part of the body, for example, a hand or an arm, is lost, the area of the brain where sensation connected to the missing limb was once recorded is "hungry for new sensations," which may now be recorded from a different part of the body, most likely one contiguous on the Penfield–Jasper diagram. In fact, he found that touching a patient's face, which is next to the arm/hand in the Sensory Homunculus, activated the "hand area" of the brain, and the brain, ignoring the physical reality, told the patient the hand was still there. Similarly, several patients had told Ramachandran that they experienced sensations of sexual arousal, up to and including orgasm, in the foot that wasn't there. The PET scans said they were right. For these patients, the area of the brain that once responded to the foot was picking up signals from an adjacent site on Penfield's map, the penis. In 2005, a "refined" sensory map from the Brain Imaging Center and Department of Neurology at the University of Hamburg in Germany found a slightly different relationship. Like Penfield's, this new map failed to include the vagina, but it did conclude that the site for the male genitals is "represented between the legs and the trunk and is thus in accord with the logical somatotrophic sequence."

But that did not invalidate Ramachandran's conclusion, summarized in his arguably inelegant, but inarguably intriguing quote from *Phantoms in the Brain: Probing the Mysteries of the Human Mind* (1999): "Maybe even many of us so-called normal people have a bit of cross-wiring, which would explain why we like to have our toes sucked."

## Building the perfect foot

With the exception of breasts, eyes, and sometimes buttocks, "small is beautiful" has always been the rule for women. In one 1996 episode of *Seinfeld*, Jerry (Jerry Seinfeld) rejects an otherwise attractive woman because she has large "man hands." In another, five years later, George (Jason Alexander) convinces another otherwise attractive woman to undergo ultimately unsuccessful surgery to reduce her prominent nose. As for large feet, they were so high a bar to romance that ladies in the Court of Louis XV used tape to make them look slimmer and smaller; other women have routinely limped along in shoes smaller than their normal size to achieve the same effect, succeeding only in pushing the bones of the foot back and to the side into the painful lump called a bunion. But no one went further than the Chinese whose glorification of the small foot ignored the Confucian prohibition against mutilating the body and for more than a thousand years encouraged women to break, crush, and bind their feet.

Some accounts say the custom originated with an empress in the Shang dynasty (1700–1027 BCE), born with a clubfoot and determined that her courtiers alter their feet to look like hers. Others blame Li Yu (937-978), the last emperor of the Southern Tang Kingdom on the eastern coast of China, so enchanted by a dancer with bound feet whose exaggerated arches were shaped like the "new moon," that he ordered all the women in his circle to do the same. Either way, bound feet became so fashionable among the upper classes that eventually even peasant women convinced themselves that binding their feet might gain them entry to a life of leisure.

In reality, the bound foot was a golden cage. Physically, it crippled its victims, thus ensuring fidelity as effectively as any medieval chastity belt. Psychologically, Freud described it as a slightly altered bow to the ever-present male fear of castration. "The Chinese custom of mutilating the female foot and them revering it like a fetish after it has been mutilated

[is]," he wrote in 1927, "as though the Chinese male wants to thank the woman for having submitted to being castrated."

For centuries, Chinese regimes up to and including the Kuomintang (Chinese Nationalists) of the 1930s and visitors like the American missionaries and British diplomats attempted to discourage foot binding. They all failed. It was not until the Communists seized power in 1949 that the practice was effectively outlawed. Unfortunately, by then, an estimated 2 billion Chinese women had bound their feet into what National Public Radio correspondent Louisa Lim has described as "the ultimate erogenous zone, with pornographic books during the Qing dynasty (1644–1912) listing forty-eight different ways of playing with women's bound feet."

The specific list of forty-eight maneuvers seems to have evaporated, at least in accessible English translation, but some sources suggest that the round opening created by bending the front of the foot down and back toward the heel served as a substitute vagina. It is verifiable that the feet were always wrapped, either because what is hidden is more erotically suggestive, or the sight of the broken and bent bones was unappealing, or the odor of the infections due to binding was seriously unromantic. In any event, the sexual intent was clear to everyone involved. In the 1980s, when American photographer Joseph Rupp, who had lived and worked in Asia, went to China to interview women for a project titled *Bound Feet* one woman told him, "I am happy to tell you about myself and foot binding, but you may not write about me or take a picture if you plan to publish them in a pornographic magazine."

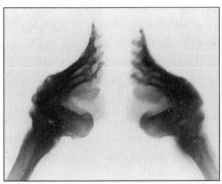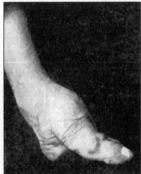

The Chinese trace their fascination with the bound female foot to one spectacular dancer or one disfigured empress. Westerners may paradoxically blame their equal fascination with a slim and dainty female extremity on the occasional open display of the female breast.

From time to time, perfectly respectable women in the West (as opposed to those women in other places that Westerners call "primitive") have worn clothing that emphasized and sometimes uncovered the breast. In Renaissance Europe, the naked breast was a common theme in the arts and so was breast-baring clothing, a fashion statement generally attributed to Agnes Sorel, the mistress of Charles VII who ruled France from 1422 to 1461. Sorel often wore dresses that left not one, but both breasts bare, knowing full well that her position ensured the sincerest form of flattery. By the seventeenth century, the fashion had crossed the Channel to normally stodgy Britain where it was adopted first by Henrietta Maria, the consort of Charles I, and later by Queen Mary II of the team of William and Mary.

Eventually, the Victorians covered the female breast and pretty much everything else. It is tempting to try to link the covering up to a need to stay warm in homes without central heating during Western Europe's "Little Ice Age," a term coined by climatologist François E. Matthes (1874–1948) in a paper published by American Geophysical Union in 1939. Unfortunately, the dates of the cold period were roughly, and with much disagreement among the scientists, set at between 1350 and 1850. After that, clearly, the Victorians were chilled by their prudery, not their climate, just as Sorel and her followers had been warmed by their nature, not their clothes.

Meanwhile, with the breast in full view, what was hidden—like the wrapped Chinese foot—was titillating. In the words of that 1930s expert on social mores par excellence, Cole Porter: "In olden days a glimpse of stocking was looked on as something shocking." His point is made in art such as Hogarth's *The Rake at Rose Tavern* (1733). That woman at the front of the painting is identifiable as a prostitute not because her breast is bare, but because she is showing her leg, her foot, her shoes, and her black silk stockings.

It is one thing to be aroused by the sight or touch or scent of a living foot attached to a living body. It is another to be attracted to the shoes that cover the foot, a fetish known as *rétifism*.

For rétifists, the shoe itself is the object of desire, to be collected, caressed, kissed, or even ejaculated into, practices that researchers in a 1998 study from Ohio State University thought may have been a form of safe sex from the sixteenth to the nineteenth centuries, a time when cases of syphilis were on the rise in Europe. This unusual love affair is named for the eighteenth century French novelist Nicolas-Edme Rétif, a.k.a. Restif or Rétif de la Bretonne (1734–1806), a contemporary of the Marquis de Sade and the man who invented the word *pornography*, from the Greek words *porno* meaning *prostitute* and *graphein* meaning *to write*. Rétif is the author

of *L'Anti-Justine* (1798), a novel he insisted was sensual rather than purely lustful like de Sade's *Justine*. Readers might find that a distinction without a difference. Amazon.com describes *L'Anti-Justine* as a "monumental odyssey of sexual depravity." As for rétifism, that arose from Rétif's real-life sighting of a young milliner in the Rue St. Denis whom he immortalized in *Le Pied de Fanchette* [*Fanchette's Foot* (1769)]. "Her foot," he wrote, "her small foot, that turns so many heads was shod with a pink pump so beautifully made and as worthy of enclosing such a beautiful foot that my eyes once fixed on that charming foot could not turn themselves away." Rétif was not the only author to delight in shoes. According to Cameron Kippen, fellow shoe fanciers have included the Roman poet Ovid, the Persian poet Omar Khayyam, and Russian novelists Leo Tolstoy and Feodor Dostoyevsky.

British psychoanalyst John Carl Flügel had a theory about that. In 1930, three years after Freud looked into the psyche to link foot fetishes to the fear of castration, Flügel looked into the mirror or, more precisely, at people looking at themselves in the mirror while dressing, to plumb the deeper meaning of clothing. What we wear, Flügel said, serves three common needs: the need for privacy ("I'm modest"), the need for protection ("I'm cold"), and the need for publicity ("Look at me!"). As for shoes, in

*The Psychology of Clothes,* published in London in 1930, Flügel agreed with Freud, and all the men who wrote the Bible. Feet, he said, served as metaphors for the male genitalia; shoes, for the vagina.

In Biblical times, the most common shoe was a flat-soled sandal. Shoes built on a thick, raised "platform" sole were worn in Asia and later in Western Europe. Butchers would wear high platform shoes while wading through offal and blood on the slaughterhouse floor, and ordinary people living in non-sewered cities such as London relied on platforms to step safely clean along the waste-strewn streets—not without consequences. The chopine, an overshoe whose sole might rise as high as 30 inches, was banned in Venice when several women miscarried after falling off their shoes while pregnant, probably as they attempted to stay dry during the annual floods in the Piazza San Marco.

The first carved out, slim, and shapely high heel may have been the one Leonardo made in 1553 as a wedding gift for Catherine of Medici (1519–1689), who, according to *Dance Magazine,* needed the extra two inches to compete with her future husband's taller mistress. After that, as the art of the period shows, both men and women wore high heels, although with some restrictions such as Louis XIV's rule that no one could wear shoes with heels higher than his and that only the nobility could wear them in red. The height limit is understandable: Every king wants to be the tallest person in the room. The ban on red shoes for the hoi-polloi is more complex. Red is the color of blood, power, fire, and the Devil. In Britain, women accused of wearing red high heels to attract a husband could be prosecuted for sorcery. By allowing his courtiers to wear red, Louis seems to have been saying that none of them was a witch. The association between witchery—or as former first lady Barbara Bush said in 1980 of Geraldine Ferraro then running as Walter Mondale's nominee for Vice President of the United States, ". . . the word that rhymes with witch"—and red shoes is still with us. On September 16, 2012, *The New York Times Book Review* illustrated its front page reviews of *The End of Men* by Hanna Rosin and *Vagina* by Naomi Wolf with a very large, very red pump whose stiletto heel was set deliberately into the back of a very small man's shoe.

Just as *Cinderella* turns the universal mother/daughter conflict into a universal fairy tale search for the rescuing prince, Hans Christian Andersen's story, *The Red Shoes* (1845), ties together the links between footwear, color, and magic. Andersen's young heroine, an orphan, naturally, covets a pair of red leather shoes. Once she has them, she decides, against all propriety, to wear them to church. As she is about to enter the door, an old

crippled soldier on a crutch leans forward to touch the shoes and com-
mands them: "Sit fast, when you dance." The command is a not a blessing;
it is a curse that fastens the shoes permanently to the girl's feet and sets
them (the shoes) dancing without pause, even when they are finally cut off
her legs, with her feet still in them. Given a pair of wooden feet, the girl
returns to church only to see her own human feet, still in their red shoes,
dancing before her. Repenting the sin of having given herself airs for the
shiny shoes, she prays for forgiveness, and her soul flies up to heaven. In
1948, Andersen's story was reinvented as the British feature film *The Red
Shoes* starring Moira Shearer as a dancer who must choose between her
lover and her starring role in a ballet based on the Andersen story. Once
again, the shoes assume control of the girl, this time carrying her forward,
over a balcony, and into the path of an oncoming train.

There is, of course, another important pair of movie red shoes, the ruby
slippers owned by the Wicked Witch of the East and appropriated by Judy
Garland's Dorothy in the *The Wizard of Oz*. In Frank Baum's original book,
the shoes were silver; for the movie, the color was changed to take advantage of the newly invented Technicolor. There were originally seven pairs;
several have disappeared, and one was stolen, but in May 2012, the pair
actually worn by Garland when she clicked her heels and went home to
Kansas was sold at auction for $2 million. For two shoes neither silver nor
ruby nor red (the camera required pink), that is witchy good indeed.

Eventually, high heels, red and otherwise, were worn not only by pros-
titutes in Paris, but also by prim Victorian ladies who liked the way they
created an arch similar to the ballerina's "banana foot," echoing a woman's
curves. Today, high heels convey power, erotic and otherwise. Long-legged
women in black net stockings and very high heels are, along with those
nurses playing their game of grown-up *Doctor*, standard erotic fare. In
*The Story of O*, the sadomasochistic novel by the pseudonymous "Pauline
Reage" (Anna Desclos, 1907–1998) published in Paris in 1954 and in the
United States nine years later, the title character—Odile—loses not only
most of her name, but also her clothes, and more importantly, her shoes.
On page one, she is "dressed as she always is, high heels, a suit with a pleat-
ed skirt, a silk blouse. . . ." On the last page, she is bereft of even the wooden
clogs she wears to her last assignation with her masters, her dignity gone
along with her last symbol of feminine power, the high-heeled shoes.

It says something about the state of pornographic literature, not to mention the state of sex in the twenty-first century, that the men in the *The
Story of O* deliver their demands to Odile in approximately 790 spoken

words (nine of which are either *whip, whipping,* or *flogging*), while Christian Grey, the eponymous title character in the more recent *Fifty Shades of Grey* (2012), submits an eleven-page contract between Dominant (him) and Submissive (her) with many clauses and subclauses, more talk about safe sex than whips and chains, and three appendixes that require the Submissive to watch her diet, work out regularly, and rate from 1 to 5 her opinion of various forms of sexual activity with the first number corresponding roughly to, "sure, why not?" and the last, "not a chance." O's agreement is a handshake among consenting adults; the contract between Dominant and Submissive resembles nothing so much as a long-winded prenup with two lawyers and a notary lurking in the background. Even if the Submissive does get to wear stiletto heels. Grey, of course.

No wonder that powerful women like Imelda Marcos, the former first lady of the Philippines, often collect high-heel shoes, in Marcos' case, an estimated 3,000 pairs, 800 of which she donated to Marikina's Footwear Museum in Manila. No wonder either, that after a decade of very flat, very expensive, very comfortable walking shoes as a fashion statement, in 1998 the girls of *Sex and the City* brought back very high, very expensive, very uncomfortable heels as standard equipment for the urban seductress.[4] And finally, absolutely no wonder that in the opening credits for *Necessary Roughness* (2011), the television show in which a female psychologist signs on to counsel male athletes, the camera does not focus on her diploma, instead closing in tight on her feet in their black, strappy, very high heels.

Marie Antoinette (1755–1793) would have understood. On June 12, 1793, proud to the end, she rode to the guillotine in an open cart, wearing, some say, two-inch heels. A few contemporary paintings and engravings do show the heeled shoes; others, have a skirt long enough to hide her feet. What is known for certain is that she was forbidden to wear a widow's black lest that draw pity from the crowd. According to an account given many years later by Rosalie Lamorlière, the shoemaker's daughter who served as maid to the imprisoned queen, Marie wore a white dress, white cap, white scarf, black stockings, and shoes made of prunella, a heavy fabric often used for clerical and academic robes. High-heeled or not, the shoes were either thrown away or thrown into the coffin with the queen and buried in an unmarked grave in a small Paris churchyard where she lay until 1815 when the coffin was dug up and the body reburied at the St. Denis Cathedral with the rest of the royals, executed and otherwise.

In March 2012, at auction in Toulan, France, a collector of Revolutionary artifacts purchased a different pair of the Marie Antoinette's shoes, size 3.5, more in line with Cinderella's than a modern American woman's. The royal shoes fetched a bid of $56,944 (43,225 Euros), $1,943,056 less than the price of the pair of Judy Garland's red shoes at auction two months later, thus proving, if proof be needed, that in life as in love what we most treasure is our fantasy—magical movie slippers, Cinderella's prettily erotic foot, and in the end as at the start, Leonardo's impossibly perfect *Vitruvian Man*.

## Notes

(1)  Roughly in order of appearance, the major monotheistic religions are Atenism, a short-lived belief system linked to the Pharaoh Amenhotep IV (Akhenaten) in the fourteenth century BCE; Judaism; Christianity; Zoroastrianism (sixth century BCE); Islam (seventh century CE); Sikhism (mid-fifteenth century); and Bahai (mid-nineteenth century).

Mormonism, founded in 1820 by the fourteen-year-old upstate New Yorker Joseph Smith and formally organized ten years later into the Church of Jesus Christ of the Latter Day Saints, considers the Trinity to be three separate beings, a situation some see as outside the Christian tradition. But Mormons, who worship only "God, the Father," consider themselves Christian and monotheistic.

Hinduism is polytheistic. Buddhism, Confucianism, Jain, Shinto, and Tao (the last two are the Japanese and Chinese words meaning "the way") are non-theistic "ways of living" based on principles of human behavior rather than veneration of one or many gods.

(2)  Count Dracula was not the first literary vampire. That honor appears to go to a spirit character in "Thalaba the Destroyer,," an epic poem by British romantic poet Robert Southey (1797). The first vampire to appear in prose seems to have been Lord Ruthven, the title character in English writer/physician John William Polidori's *The Vampyre*, a novella/short story written while Polidori was on holiday in Switzerland with Mary Shelley and Byron during the summer of 1816 that also produced the much better known *Frankenstein*. Polidori's story was published in London in the April 1819 edition of *The New Monthly Magazine*, which erroneously attributed authorship to Byron.

(3)  The literal meaning of homunculus is little man; historically, the word was used to describe the completely formed, very small human body once believed to be contained in the spermatozoa.

(4)  Testifying to the popularity—but not the comfort—of one brand of stilettos featured prominently on the television show, in August 2012 the designer's store on Madison Avenue in New York was robbed of thirty-four handbags, each priced at more than $2,000, but only one pair of shoes. Clearly the thieves were seeking chic without pain, the promise of "Dr. Scholl's" For Her High Heel Insoles," introduced in 2011 with a "unique design [that] helps prevent foot aches and pains caused by high heels two inches and higher."

# APPENDIX

# PAIS, PED, POD

"When I use a word,"
Humpty Dumpty said, in rather a scornful tone,
"it means just what I choose it to mean—neither more nor less."

Lewis Carroll, *Through the Looking Glass* (1871)

QUESTION: When is a *ped* not a foot?

ANSWER: When its etymology tracks back not to *pes,* the Latin word for *foot,* but to *pais,* the Greek word for *child.*

For example, the verb *pedal,* from the Latin *pes,* means to move something with your foot, and the noun *pedal* is the thing you move with your foot to move something else, say, a bicycle. But a *pedagogue,* from the Greek *pais* plus the add-on *-agogos* (which means *to lead),* is a *person who teaches.*

These rules apply, of course, regardless of where the *p* letters sit in any given word. Although they are most commonly up front, they may also be imbedded in the middle as in encyclo*ped*ia, ex*ped*ite, and im*ped*iment.

The first, *encyclopedia,* is a *pais* word from the Greek compound *enkyklios,* whose literal meaning is *in* [*en*] a *circle* [*kyklos*], but the word may also mean *general.* Adding *paideia/pedia,* which means *education* or *child-rearing,* gives you "general education," which is pretty much what an encyclopedia is. Our use of the word encyclopedia as the name for a reference book dates to the eighteenth century and Denis Diderot's *Encyclopédie ou Dictionnaire Raisonné des Sciences, des Arts, et des Métiers,* published in a series *of* volumes over twenty or twenty-five years after 1751. Its (temporary) suppression by the French church and state, was one of the matches that lit the fires of the Revolution. Anne-Robert-Jacques Turgot, Baron de

158

l'Aulne, an economist who had served as controller general under Louise XVI, attempted to convince the King to put in place financial reforms to even the field for French society. Turgot was sensible but completely unsuccessful; he died eight years before the deluge he predicted arrived to wash away the *Ancien Regime:*. The Scots, being a people whose heads (and climate) were cooler than the French, produced the second major encyclopedia, *The Encyclopaedia Britannica,* in Edinburgh in 1768 during the period known as the Scottish Enlightenment. Its publication went off without anyone's spilling a single drop of blood.

*Expedite,* the second word with *–ped* in the middle, comes from the Latin verb *expedire,* which means to *free the feet from fetters,* in other words, to speed someone or more likely something, on its way. The third, *impediment,* is the exact opposite, from the Latin verb *impedire,* which means *to shackle the feet.*

As for the concluding *peds* and *pedes* in words such as bi*ped,* quadru*ped,* centi*pede* and milli*pede,* they simply and directly join the Latin prefixes for two, four, one hundred, and one thousand with *feet* to describe creatures with a certain number of legs. I do not know why one set ends with *-ped* and the other with *–pede.* Maybe they simply sound better that way, but they may sound better that way because that's how we're used to hearing them sound. What a pity there is no *-ped* word for mystery, which is what the word *stampede* presents. A stampede is an impulsive, rushed run of frightened animals or people. Because they all run on their feet, you'd think the word has a *–ped* connection. My dictionaries say no, and so does the *Online Etymology Dictionary,* which credits it to the "Mex.Sp. *estampida,* from Sp., 'an uproar,' from *estamper* 'to stamp, press, pound,' from the same Germanic root that yielded English stamp (v.)."

*Pod* can also be a puzzle. Obviously, *podiatrist* comes from the Greek *pous* meaning *foot* plus *iatros* meaning *physician. Podagra,* from the Latin *pod* or the Greek *pous* plus the Greek *agra* meaning *to catch,* translates loosely to *foot trap,* so it is the perfectly sensible medical name for *gout.* As for the adjective *podgy,* etymology links that to *pudgy* which may be related to *pudsy,* an eighteenth century term meaning *plump* derived from *pud,* meaning *hand* or *forepaw.* Bingo.

But which early nineteenth century American seaman, whaler, naturalist, or just plain citizen first used the word *pod* to describe a small group of animals, most commonly whales or seals, and why he (or she) chose to do so is lost to history. So is how it came to mean, as my usually reliable *Webster's Seventh Collegiate Dictionary*—the one I found upstairs at Manhattan's Argosy Bookstore when the earlier edition that got me through college fell apart in my hands—notes, a "bit socket in a brace or a straight groove or channel in the barrel of an auger." *Podzol*—sometimes spelled *podsol*—is a complete mystery. Unless, that is, you speak Russian in which case you would know that *pod* means *under,* *zol* means *ash,* and when you put the two together you get a word that means *leached soil,* the pale dirt from which soluble particles have been washed away. And, except for the fact that you stand on the ground, the word has nothing at all to do with your feet.

Last but not least, there are words that begin with the letter "p," say something about your feet, but have no relationship to either *pes* or *pais* or *pod.* One good example is *pace,* which can be either a noun (a step) or a verb (to step). This one comes from *passus,* the Latin word for *step.*

How to identify the root of any ped, pais, pod word and therefore decipher its meaning? You could guess. Or you could leaf through one of the *Webster's,* or you can click onto the excellent *Online Etymology Dictionary* whose list of sixty-seven sources includes dictionaries in Arabic, French, German, Greek, Icelandic, Latin, Nahuatl (the language of the Aztecs), Norse, Persian, Sanskrit, and Slang—American, British, and Buckish, the last a nineteenth century British word meaning *dandified* or *foppish,* which in turn means excessively refined and fastidious.

From time to time, words disappear from our collective vocabulary, so for special ones like *podocarp,* which seems to be gone from contemporary dictionaries for lay audiences, there's always the 3,210-page, 16-pound, 1941 edition of the *Webster's New International Dictionary,* which defines *podocarp* as the stemmed (i.e., *footed*) fruit (from the Greek word *karpus*) of an evergreen tree or shrub belonging to the genus *Podocarpus.*

Identifying these differences in *pais, ped,* and *pod* is a lovely parlor game, but language itself is not trivial. Ralph Waldo Emerson called it "the archives of history." Jonathan Swift wrote that "proper words in their proper places make the true definition of a style," and William James, that they "make a difference in our moral life." Lewis Carroll's sly dismissal of linguistic standards is a warning to use words carefully to say what we truly mean, and truly mean what we say lest reality crack into pieces like Humpty Dumpty himself.

Our feet are unique, but so is our language, distinguishing us from all the creatures of the earth. Birds sing, wolves howl, and even the lowly male cricket chirps (although with his wings, not a voice box). Dogs bark and cats purr and both position their ears to sends specific signals, one to another. But only human beings have the physical and intellectual equip-ment required to speak their words and thoughts.

And to complete the link between foot and tongue—the language, of course, not the muscles in your mouth—in 2011, a group of psychologists at the University of Auckland in New Zealand reported in the journals *Science* and *Nature* that, based on their analysis of languages from every corner of the globe, we spoke up first where we stood up first, in Africa where Raymond Dart found his wonderful Taung Child.

# SOURCES AND BIBLIOGRAPHY

Note: The Internet sources cited in this bibliography were accessed online between September 2011 and October 2012.

## 1 • Destiny

Abedon, Stephen T., "Descent of Man, Supplemental Lecture," updated May, 4, 1997. http://www.mansfield.ohio-state.edu/~sabedon/biol1530.htm

Ahern, J.C., "Foramen Magnum Position Variation in Pan troglodytes, Plio-Pleistocene Hominids, and Recent Homo Sapiens: Implications for Recognizing the Earliest Hominids." *American Journal of Physical Anthropology*, July 2005, 127(3):267–76. http://www.ncbi.nlm.nih.gov/pubmed/15558606

"Analysis of Early Hominins." *Early Hominin Evolution*, updated on December 23, 2011. http://anthro.palomar.edu/hominid/australo_2.htm

"Anatomical Evidence for Bipedalism." *eFossils*, updated January 23, 2012. http://www.efossils.org/book/step-step-evolution-bipedalism

"Australopithecus africanus." *ArcheologyInfo.com.* http://archaeologyinfo.com/australopithecus-africanus/

"Ask the Experts." *Scientific American*, October 21, 1999. http://www.scientificamerican.com/article.cfm?id=it-seems-that-in-almost-a

Berman, David S., Robert R. Reisz, Diane Scott, A.C. Henrici, Stuart S. Sumida, and Thomas Martens, "Early Permian Reptile." *Science*, November 3, 2000, 290: 969–72.

"Biographies: Raymond Dart." *Talk.origins.* http://www.talkorigins.org/faqs/homs/rdart.html

"Bipedalism." http://www.sccs.swarthmore.edu/users/08/ajb/tmve/wiki100k/docs/Bipedalism.html

Blumenschine, Robert J., and John A. Cavallo, "Scavenging and Human Evolution." http://www.mesacc.edu/dept/d10/asb/origins/hominid_journey/scavenging.

Brain, C.K., "Raymond Dart and Our African origins," from *Laura Garwin and Tim Lincoln, editors, A Century of Nature: Twenty-One Discoveries that Changed Science and the World,* 2003. http://www.press.uchicago.edu/Misc/Chicago/284158_brain.html

"Coastal People." http://www.riverapes.com/AAH/Hardy/Hardy.htm

Elftman, Herbert, and John Manter, "The Evolution of the Human Foot, with Especial Reference to the Joints." *Journal of Anatomy,* October 1935, 70(Pt 1):56–67. http://www.ncbi.nlm.nih.gov/pmc/articles/PMC1249279/pdf/janat00534-0078.pdf

"Evolution of Human 'Super-Brain' Tied to Development of Bipedalism, Tool-Making." *Science Daily,* April 20, 2011. http://www.sciencedaily.com/releases/2011/04/110420125510.htm

"Footprints Can Help Identify People Just Like Finger Prints," The Indian Express, updated June 6, 2012. *http://www.indianexpress.com/news/footprints-can-help-identify-people-just-lik/844033/*

Grenier, Thomas M., "Re: How Did Human Foot Evolve," *MadSciNetwork: Evolution,* April 21, 2000. http://www.madsci.org/posts/archives/2000-04/956600859.Ev.r.html

Grigg, Russell, "Raymond Dart and the 'Missing Link.' " *Creationministries.com.* http://creation.com/raymond-dart-and-the-missing-link

Harcourt-Smith, W.E.H, and L.C. Aiello, "Fossils, Feet and the Evolution of Human Bipedal Locomotion." *Journal of Anatomy,* May 2004, 204(5): 403–16. http://www.ncbi.nlm.nih.gov/pmc/articles/PMC1571304/

Heaton, Mrs. Charles W., *Leonardo da Vinci and His Works,* Kessinger Publishing, Whitefish, MT, 2004.

Hendrickson, Robert, *The Facts on File Encyclopedia of Word and Phrase Origins,* 4th edition, Checkmark Books, New York, 2008).

"Hominin Hunting." *Smithsonian.com,* October 17, 2011. http://blogs.smithsonianmag.com/hominids/2011/10/how-africa-became-the-cradle-of-humankind/

Kemp, Martin, *Leonardo Seen From the Inside Out.* Oxford University Press, New York, 2004.

Lieberman, Daniel E., "Those feet in ancient times," *Nature,* March 29, 2012, 483: 550–1, http://www.nature.com/nature/journal/v483/n7391/full/483550a.html

Lighter, J.E., editor, *Random House Historical Dictionary of American Slang,* volume 1, Random House, New York, 1994.

"New View of Human Evolution? 3.2 Million-Year-Old Fossil Foot Bone Supports Humanlike Bipedalism in Lucy's Species," *Science Daily.com,* February 10, 2011. http://www.sciencedaily.com/releases/2011/02/110210141215.htm

O'Neil, Dennis, "Evolution of Modern Humans: A Survey of the Biological and Cultural Evolution of Archaic and Modern Homo Sapiens," 2009–2012. http://anthro.palomar.edu/homo2/mod_homo_2.htm

"Primate." *Encyclopædia Britannica Online,* June 4, 2012. http://www.britannica.com/EBchecked/topic/476264/primate/225207/Vertebral-column-and-posture.

"Raymond Dart." *StrangeScience.com*. http://www.strangescience.net/dart.htm

"Siegfried Woldhek: How He Found the True Face of Leonardo." *TED conferences*, filmed February 2008, posted April 2008. http://www.ted.com/talks/siegfried_woldhek_shows_how_he_found_the_true_face_of_leonardo.html

Stephan, Annelisa, "Leonardo da Vinci, Anatomist." *Getty.edu*, June 4, 2010. http://blogs.getty.edu/iris/leonardo-da-vinci-anatomist/

Viegas, Jennifer, "Footprints Show How Our Ancestors Walked." *Discovery News*, July 19, 2011. http://news.discovery.com/human/ancient-footprint-ancestor-walked-upright-1107019.html

Watson, James and John McClelland, "Distinguishing Human From Animal Bone." *University of Arizona/Arizona State Museum*. http://www.statemuseum.arizona.edu/crservices/human_animal_bone.shtml

Wilford, John Noble, "Some Prehumans Feasted on Bark Instead of Grasses." *The New York Times*, June 28, 2012.

## 2  •  Disability

Adams, John Paul, "The Twelve Tables, 451–450 B.C.." June 10, 2009. http://www.csun.edu/~hcfll004/12tables.html

Alvarado, David M., Hyuliya Aferol, Kevin McCall, Jason B. Huang, Matthew Techy, Jillian Buchan, Janet Cady, Patrick R. Gonzales, Matthew B. Dobbs, and Christina A. Gurnett1, "Familial Isolated Clubfoot Is Associated with Recurrent Chromosome 17q23.1q23.2 Microduplications Containing TBX4." *American Journal of Human Genetics*, July 9, 2010, 87(1): 154–60. http://www.cell.com/AJHG/abstract/S0002-9297(10)00313-7

Alvarado, David M., Kevin McCall, Hyuliya Aferol, Matthew J. Silva, Joel R. Garbow, William M. Spees, Tarpit Patel, Marilyn Siegel, Matthew B. Dobbs, and Christina A. Gurnett, "Pitx1 Haploinsufficiency Causes Clubfoot in Humans and a Clubfoot-like Phenotype in Mice." Human Molecular Genetics, published online July 20, 2011. http://hmg.oxfordjournals.org/content/early/2011/07/20/hmg.ddr313

Alvares, Gonzalo, Francisco C. Ceballos, and Celsa Quinteiro, "The Role of Inbreeding in the Extinction of a European Royal Dynasty." *PLoS ONE.com*, April 15, 2009, 4(4): e5174. http://www.plosone.org/article/info:doi/10.1371/journal.pone.0005174

"Baillie, Matthew," Britannica Online. http://www.eb.com:180/cgi-bin/g?DocF=micro/46/71.html

Bates, A.W., "The *De monstrorum* of Fortunio Liceti: a landmark of descriptive teratology." *Journal of Medical Biography*, 201, 9: 49–54.

Bazopoulou-Kyrkanidou, E. "What makes Hephaestus lame?" *American Journal of Medical Genetics*, October 1997, 72(2):144–55. http://www.cc.columbia.edu/cu/cup/

Blumberg, Roger B., "Notes, About Mendel's paper and the English translation at MendelWeb." http://www.mendelweb.org/MWNotes.html

Brignell, Victoria, "Ancient world." *The New Statesman*, April 7, 2008. http://www.newstatesman.com/blogs/crips-column/2008/04/disabled-slaves-child-roman

Bristow, Jennie, "It is right that society offers late abortions." *Spiked.com*, May 5, 2011. http://www.spiked-online.com/index.php/site/printable/10483/

Burrus, Trevor, "One Generation of Oliver Wendell Holmes, Jr. Is Enough," *Cato @ Liberty*, June 23, 2011. http://www.cato-at-liberty.org/one-generation-of-oliver-wendell-holmes-jr-is-enough/

Buscaglia, Marino and Denis Duboule, "Developmental Biology in Geneva: A Three Century-long Tradition." International Journal of Developmental Biology, 2002, 46:5–13. http://www.ijdb.ehu.es/web/contents.php?vol=46&issue=1&doi=11902688

"Clubfoot—No Clear Cause, But Possibly a Cure." *AlertNet.com*, January 25, 2011. http://www.trust.org/alertnet/news/health-clubfoot-no-clear-cause-but-possibly-a-cure/

Crites, James A., "Chinese Foot Binding." James @eCrites.com, October 25, 1995. http://www.angelfire.com/ca/beekeeper/foot.html

Corner, George W., "Congenital Malformations: The Problems and the Task." *First International Conference on Congenital Malformations.* J.B. Lippincott, Philadelphia, 1960.

Dobbs, Matthew B., José A. Morcuende, Christina A .Gurnett, and Ignacio V. Ponseti, "Treatment of Idiopathic Clubfoot, An Historical Review." *Iowa Orthopaedic Journal*, 2000, 20: 59–64. http://www.ncbi.nlm.nih.gov/pmc/articles/PMC1888755/

Dobson, Roger, "Infection rate after some surgery types in England is triple that in US." British Medical Journal, May 14, 2005, 330(7500): 1104. http://www.ncbi.nlm.nih.gov/pmc/articles/PMC557928/

"Dry As Dust, A Fortean in the Archives: Erotic Secrets of Lord Byron's Tomb." October 16, 2010. http://blogs.forteana.org/mike

Ehrenfried, Albert, "Clubfoot. A statistical note." *Journal of Bone and Joint Surgery*, 1914. http://www.jbjs.org/data/Journals/JBJS/1002/644.pdf

"Ensoulment." *Absolute Astronomy.com*. http://www.absoluteastronomy.com/topics/Ensoulment

"Ephesus." *Bible History On line*. http://www.bible-history.com/maps/romanempire/Ephesus.html

*The Etymologies of Isadore of Seville, Liber XI.* https://sites.google.com/site/theetymologies/complete-text/liber-xi

"Eugenics Organizations." Dolan DNA Learning Center, Cold Spring Harbor Laboratory, Cold Spring Harbor, NY. http://www.eugenicsarchive.org/html/eugenics/static/themes/19.html

"First Gene for Clubfoot Identified." *Science Daily*, October 23, 2008. http://www.sciencedaily.com/releases/2008/10/081023144101.htm

First International "Fortunio Liceti." *Museo Galileo*, 2008. http://brunelleschi.imiss.fi.iy/itineraries/biography/FotunioLiceti.html

"Footprints of the Buddha, Buddhist Studies: Symbols/Iconography." *BDEA/BuddhaNet*. http://www.buddhanet.net/e-learning/history/b_feet.htm

Gilbert, Martin, "Churchill and Eugenics." *The Churchill Centre and Museum*, May 31, 2009. http://www.winstonchurchill.org/support/the-churchill-centre/publications/finest-hour-online/594-churchill-and-eugenics

Gurnett1, Christina A., Farhang Alaee, Lisa M. Kruse, David M. Desruisseau, Jacqueline T. Hecht, Carol A. Wise, Anne M. Bowcock, and Matthew B. Dobbs, "Asymmetric Lower-Limb Malformations in Individuals with Homeobox PITX1 Gene Mutation." *American Journal of Human Genetics*, November 17, 2008, 83(5): 616–622. http://www.ncbi.nlm.nih.gov/pmc/articles/PMC2668044/

Harris, W.V., "Child-Exposure in the Roman Empire," JSTOR: The Journal of Roman Studies, 1994, 84: 1–22. http://www.jstor.org/pss/300867

Head, Tom, "Forced Sterilization in the United States." *About.com: Civil Liberties*. http://civilliberty.about.com/od/gendersexuality/tp/Forced-Sterilization-History.htm

Hermodorus, *Dictionary of Greek and Roman Biography and Mythology*. http://www.ancientlibrary.com/smith-bio/1527.html

"The History of Attitudes to Disabled People." The British Film Institute. http://www.bfi.org.uk/education/teaching/disability/thinking/

Hortman, Wm. Jeffreys, "The Vulnerable Hero: Byron and Modern Athletics." Paper presented at the *Missolonghi Byron Conference*, May 16–24, 2004. http://rea.teimes.gr/byronlib/media/files/paper_htm/The_Vulnerable_Hero.htm

"How Keeping It in the Family Spelled the End of the Line for the Hapsburg Royal Dynasty." *The Daily Mail Online*, updated April 15, 2009. http://www.dailymail.co.uk/sciencetech/article-1170143/How-keeping-family-spelled-end-line-inbred-royal-dynasty.html#ixzz1wyAnDBT1

"Impact Revealed of Having Child with Clubfoot." Office of External Affairs, University of Aberdeen, King's College, Aberdeen, May 27, 2011. http://www.abdn.ac.uk/news/archive-details-10244.php

"King Tut's Mom and Dad ID'd." *Live Science*, February 16, 2010. http://www.livescience.com/8092-king-tut-mom-dad-id-ed.html

Lienhard, John H., "No. 603: Pare's Monsters," Engines of Our Ingenuity. http://www.uh.edu/engines/epi603.htm

Lister, Joseph (1827–1912). *BLTC Research*. http://www.general-anaesthesia.com/people/joseph-lister.html

Lombardo, Paul A. "The American Breed: Nazi Eugenics and the Origins of the

Pioneer Fund." *Albany Law Review*, 65(3): 822. Cited in Harry H. Laughlin, *Wikipedia*. http://en.wikipedia.org/wiki/Harry_H._Laughlin

Lombardo, Paul A. "Eugenic Sterilization Laws." Image Archive on the American Eugenics Movement, Dolan DNSA Learning Center, Cold Spring Harbor Laboratory, Cold Spring Harbor, New York. http://www.eugenicsarchive.org/html/eugenics/essay8text.html

Lord, Janet E., "Disabilities, People with." *eNotes*, 2005. http://www.enotes.com/disabilities-people-with-reference/disabilities-people-with

Macmillan, Malcolm, "The Case of Bridey Murphy," University of Melbourne and Deakin University, updated 10/28/2011. http://socrates.berkeley.edu/~kihlstrm/BrideyMurphy/BrideyMurphyIndex.htm

"Maternal Impressions and Reincarnation Theory." *Anthro Doula.com*, May 25, 2010. http://anthrodoula.blogspot.com/2010/05/maternal-impressions-and-reincarnation.html

Moran, Diane, "Infanticide." *Encyclopedia of Death and Dying*, Routledge, New York, November 9, 2001. http://www.deathreference.com/Ho-Ka/Infanticide.html#ixzz1bucioMHr

MRC Vitamin Study Research Group, "Prevention of Neural Tube Defects: Results of the Medical Research Council Vitamin Study." *The Lancet*, 1991, 338: 131. http://www.mrc.ac.uk/Achievementsimpact/Storiesofimpact/Folicacid/index.htm

"New Kingdom of Ancient Egypt, From N.S. Gill's Ancient/Classical History Glossary." *About.com*. http://ancienthistory.about.com/cs/egypt/g/newkingdom.htm

Noel, Roden, *Life of Lord Byron*, 1890. Cited at http://www.archive.org/stream/lifeoflordbyron00noeluoft/lifeoflordbyron00noeluoft_djvu.txt

Nordin, S., M. Aidura, L. Razak, and WI. Faisham, "Controversies in Congenital Clubfoot: Literature Review." *Malaysian Journal of Medical Science*, January 2002, 9(1): 34–40.

Perry, David L., "Abortion and Personhood: Historical and Comparative Notes." http://home.earthlink.net/~davidlperry/abortion.htm

Pietrucin-Materek, Marta, Edwin R. van Teijlingen, Simon Barker, Karen Forrest, and Zosia Miedzybrodzka, "Parenting a Child with Clubfoot: A Qualitative Study." *International Journal of Orthopaedic and Trauma Nursing*, November 2011, 1(4): 176–84. http://www.orthopaedictraumanursing.com/article/S1878-1241(11)00006-2/abstract

"PITX1 paired-like homeodomain 1 [ Homo sapiens ] /Gene ID: 5307, updated on 7-Nov-2011 Summary Official Symbol /PITX1 provided by HGNC." National Institutes of Health. http://www.ncbi.nlm.nih.gov/gene/5307

"Polydactyly and syndactyly." *Answers.com*. http://www.answers.com/topic/polydactyly-and-syndactyly#ixzz1bok1iFDc/

Powell, John, "The Clevedon Ventilator and poliomyelitis." http://www.johnpowell.net/pages/clevedon.htm

"St. Paul's Epistle to the Ephesians" by Gore. Free
    Fiction Books. http://www.freefictionbooks.org/
    books/s/21320-st-pauls-epistle-to-the-ephesians-by-gore?start=88

Schemm, Paul, "Studies Reveal King Tut's Sad Life and Death." *Science
    on MSNBC.com*, updated February 10, 2010. hhttp://www.msnbc.
    msn.com/id/35423552/ns/technology_and_science-science/t/
    studies-reveal-king-tuts-sad-life-death/#

"Section I: Poliomyelitis, Treatment, and Prevention Prior to 1955," *VaccineEthics.
    org*, http://www.vaccineethics.org/salk_polio/pre1955_B.php

Severson, Kim, "Payments for Victims of Eugenics Are Shelved." *The New York
    Times*, June 21, 2012.

Showell, C., O. Binder, and F.L. Conlon, "T-box genes in early embryogenesis."
    *Developmental Dynamics*, January 2004, 229(1):201–18. http://www.ncbi.
    nlm.nih.gov/pubmed/14699590

Singhal, Hemant, "Wound Infection." *eMedicine.com*. http://emedicine.medscape.
    com/article/188988-overview

"Sir Hector Cameron," *The University of Glasgow story*. http://www.
    universitystory.gla.ac.uk/biography/?id=WH1419&type=P

Smith, J., "T-Box Genes: What They Do and How They Do It." *Trends in Genetics*,
    April 1999, 15(4):154–8. http://www.ncbi.nlm.nih.gov/pubmed/10203826

Smith, Peter, "Louis Stromeyer (1804–76): German orthopaedic and military
    surgeon and his links with Britain." *Journal of Medical Biography*, May 2006,
    14(2):65–74. http://jmb.rsmjournals.com/content/14/2/65.short

Stevenson, Ian, "A New Look at Maternal Impressions: An Analysis of 50
    Published Cases and Reports of Two Recent Examples." Journal of Scientific
    Exploration, 1992, 6(4), 353–73.

Szeto, Daniel P., Concepción Rodriguez-Esteban, Aimee K. Ryan, Shawn M.
    O'Connell, Forrest Liu, Chrissa Kioussi, Anatoli S. Gleiberman, Juan Carlos
    Izpisúa-Belmonte, and Michael G. Rosenfeld, "Role of the Bicoid-Related
    Homeodomain Factor Pitx1 in Specifying Hindlimb Morphogenesis and
    Pituitary Development." *Genes and Development*, February 15, 1999,
    13(4):484–94. http://genesdev.cshlp.org/content/13/4/484.full

Temtamy, Samia, and Mona Aglan, "Consanguinity and Genetic Diseases
    in Egypt." *Middle East Journal of Medical Genetics*, January 2012, 1(1):
    12–17. http://journals.lww.com/mejmedgen/Abstract/2012/01000/
    Consanguinity_and_genetic_disorders_in_Egypt.3.aspx

Turpin, Raymond, and Jerome LeJeune, "Congenital Human Anomalies Due to
    Chromosome Aberrations," *Maandschr Kindergeneeskd*, May 1961, 29:149–
    73. http://www.ncbi.nlm.nih.gov/pubmed/13778772

"Twenty Medical Classics of the Jefferson Era." *University of Virginia Health
    System*. http://www.hsl.virginia.edu/historical/rare_books/classics/#Scarpa

Wernick, Robert, "Prince Tallyrand, The Ultimate Survivor." 2007. http://www.
    robertwernick.com/articles/talleyrand.htm

Velasquez, Leticia, "Ominous Debate in France over
  Mandatory Pre-Natal Genetic Testing." *Spero News*,
  May 24, 2011. http://www.speroforum.com/a/54340/
  Ominous-debate-in-France-over-mandatory-prenatal-genetic-testing

Zheng, Zhengui, and Martin J. Cohn, "Developmental Basis of Sexually
  Dimorphic Digit Ratios. PNAS, September 27, 2011, 108 (39): 16289–16294.
  http://www.pnas.org/content/108/39/16289

## 3 • Difference

Abdel-Fattah, M.M., M.M. Hassanin, F.A. Felembane, and M.T. Nassaane,
  "Flatfoot among Saudi Arabian Army Recruits: Prevalence and Risk Factors."
  *Eastern Mediterranean Health Journal*, January–March 2006, 12(1&2). http://
  www.irdpq.qc.ca/communication/publications/PDF/Part%205%20Foot.pdf

"About Dr. Scholl." Funding Universe. http://www.fundinguniverse.com/
  company-histories/SSL-International-plc-Company-History.html

"Adult (Acquired) Flatfoot." *Orthoinfo.com*, reviewed December 2011. http://
  orthoinfo.aaos.org/topic.cfm?topic=a00173

"The Agony of Pointe Work." *theperfectpointe.com*. http://www.the-perfect-
  pointe.com/Agony.html

Alexander, Harriet, "Archless, but they're no clodhoppers." *SMH.com.au*, July 14,
  2005. http://www.smh.com.au/news/national/archless-but-theyre-no-clodho
  ppers/2005/07/13/1120934304116.html

Armstrong, Lois, "Dr. Charles Lowman at Age 95 Is Still Straightening Out
  Kids." *People,* April 28, 1975. http://www.people.com/people/archive/
  article/0,,20065185,00.html

Blandin, N., P.J. Parquet, and D. Bailly, "Separation Anxiety. Theoretical
  Considerations." *il Encephale,* March–April 1994, 20(2):121–9. http://www.
  ncbi.nlm.nih.gov/pubmed/8050378

"Buck Dancing and Flatfooting | Appalachian Dancing." *Squidoo.com*. http://
  www.squidoo.com/buck-dancing-flat-footing

Burke, Gerald L., "The Surgical Treatment of Flatfeet." *Canadian Medical
  Association Journal*, October 1940, Canadian Medical Association Journal,
  October 1940, 43(4): 327–31. http://www.ncbi.nlm.nih.gov/pmc/articles/
  PMC1826497/

Castro-Aragon O, S. Vallurupalli, M. Warner, V. Panchbhavi, and S. Trevino,
  "Ethnic radiographic foot differences." *Foot & Ankle International*, January
  2009, 30(1):57–61. http://www.ncbi.nlm.nih.gov/pubmed/19176187

"Charles LeRoy Lowman (1879–1977) Orthopedic Surgeon." April 14, 2010.
  http://zagria.blogspot.com/2010/04/charles-leroy-lowman-1879-1977.
  html#ixzz1iWKGtowp

Cherry, Kendra, "John Bowlby Biography (1907–1990)." *About.com. Psychology.*
  http://psychology.about.com/od/profilesal/p/john-bowlby.htm

Cherry, Kendra, "What is Phrenology?" *About.com.Psychology*. http://psychology. about.com/od/historyofpsychology/f/phrenology.htm

Clapton, Jayne, and Jennifer Fitzgerald, "The History of Disability: A History of 'Otherness.'" *New Renaissance Magazine*, May 17, 2012. http://www.ru.org

The Cleveland Clinic, "Benign Hypermobility Joint Syndrome." http://www.cchs. net/health/health-info/docs/1700/1722.asp?index=3971

"The Diseases and Infirmities Exempting from the Draft," The New York Times, November 15, 1863. http://www.nytimes.com/1863/11/15/news/the-diseases-and-infirmities-exempting-from-the-draft.html

"Every Contact Leaves a Trace." http://www.southwalespolicemuseum.org.uk/en/ content/cms/visit_the_archives/history_of_fingerpri/history_of_fingerpri. aspx

"Flatfeet." University of Maryland Medical Center. http://www.umm.edu/ency/ article/001262.htm

"Flatfoot," *The Random House Historical Dictionary of American Slang, Volume 1, A–G*. Random House, New York, 1994.

"Flatfoot," Word of the Day, Merriam Webster.com, September 9, 2011. http:// www.merriam-webster.com/word-of-the-day/2011/09/09/

"Flatfoot Floosie (with the Floy Floy)," Podiatry Arena. December 16, 2010. http://www.podiatry-arena.com/podiatry-forum/showthread.php?t=59107

"Frequency of Foot Disorders Differs Between African-Americans and Whites," *Science News*, November 8, 2010. http://www.sciencedaily.com/ releases/2010/11/101108151420.htm

Fritscher, Lisa, "Xenophobia. Fear of Strangers." About.com. Updated December 23, 2009. http://phobias.about.com/od/phobiaslist/a/xenophobia.htm

Gray, Henry, *Anatomy of the Human Body, The Unabridged Running Press Edition Of The American Classic*. Running Press, Philadelphia, 1974.

"Gumshoe." *Words@Random*. http://www.randomhouse.com/wotd/index. pperl?date=19981110

Hayward, Anthony, "Charles S Dubin: Television Director Who Overcame Being Blacklisted and Went On to Make 44 episodes of 'M*A*S*H.'" *The Independent*, September 19, 2001. http://www.independent.co.uk/news/ obituaries/charles-s-dubin-television-director-who-overcame-being-blacklisted-and-went-on-to-make-44-episodes-of-mash-2356804.html

Hearne, Kevin, "The Demonization of Pan." 1998. http://www.mesacc. edu/~tomshoemaker/StudentPapers/pan.html

"High Insteps, High Arches: Ballet Feet." March 31, 2009. http://apricot. wordpress.com/2009/03/31/high-insteps-high-arches-ballet-feet/

"History of Foot Orthotics." *FastTech*, 2012. http://www.fastechlabs.net/ brief-history

H.R. 257: Infant Protection and Baby Switching Prevention Act of 2007, 110th

Congress: 2007–2008. http://www.overcriminalized.com/LegislationDetail. aspx?id=337

Inches to U.S. Men's Shoe Sizes–Conversion Chart. http://shoes.about.com/od/ fitcomfort/a/men_inches.htm

Ivry, Benjamin, "TV Director Charles Dubin's Sweet Revenge." *The Jewish Daily Forward*, December 2, 2011. http://blogs.forward.com/ the-arty-semite/147263/#ixzz1gwjzpVgr,

"Jane Seymour." *FriendsofJane.com*. http://www.friendsofjane.com/aaj_artsed. html

"Judgement: Streicher." The Avalon Project. http://avalon.law.yale.edu/imt/ judstrei.asp

Keds, Our Story. http://www.keds.com/store/SiteController/keds/ourstorypage

Kippen, Cameron, "A Potted History of Podiatry." *Foot Talk*, December 22, 2008, http://foottalk.blogspot.com/2008/12/potted-history-of-podiatry.html

Kolata, Gina, "Close Look at Orthotics Raises a Welter of Doubts." *New York Times,* January 7, 2011.

Leahy, Maureen, "Treatment for pediatric pes planus debated." *AAOS Now,* May 2011. http://www.aaos.org/news/aaosnow/may11/clinical7.asp

Lewis, Jack, "Lead Poisoning: A Historical Perspective." *EPA Journal*, May 1985. http://www.epa.gov/history/topics/perspect/lead.html

Malani, Preeti N., "Harrison's Principle of Internal Medicine," Book and Media Reviews, *JAMA*, 308 (17): 1813–14. November 7, 2012. http://jama. jamanetwork.com/article.aspx?articleid=1389591

McLeod, S. A. "Simply Psychology; Attachment Theory." 2009. http://www. simplypsychology.org/attachment.html

McNally, Shelagh, "Flatfeet: The Achilles Heel of Middle Age." *Knee1.com*, November 11, 2005. http://www.knee1.com/News/feature_mainstory. cfm/294

"Medicinal Animal Experimentation: Pointless Cruelty or Necessary Evil." http:// medicinalanimalexperimentationpointlesscrueltyornecessaryevil.weebly. com/history.html

"New View of Human Evolution? 3.2 Million-Year-Old Fossil Foot Bone Supports Humanlike Bipedalism in Lucy's Species." *Science Daily*, February 10, 2011. http://www.sciencedaily.com/releases/2011/02/110210141215.htm

"The Other." February 5, 2009. http://academic.brooklyn.cuny.edu/english/ melani/cs6/other.html

"Patients Spared Severe Arthritis, Ankle Fusion Over Nine-Year Follow Up." July 8, 2010. http://www.hss.edu/newsroom_new-surgery-improves-outcomes- for-severe-flat-foot-deformity.asp

"Pes Planus (Flexible Flatfoot)." Children's Memorial Hospital. Reviewed June 2010. http://www.childrensmemorial.org/depts/orthopaedic/pes-plano.aspx

"Physical Characteristics of the Buddha." eNotes. http://www.enotes.com/topic/
    Physical_characteristics_of_the_Buddha

"Negrofoot." Podunk The Book.com. 2011. http://www.podunkthebook.com/
    negro-foot-virginia.html

Polsdorfer, Ricker, "Flatfoot (Pes Planus; Pes Planovalgus; 'Fallen 'Arches')',"
    reviewed September 2011 by John C. Keel. http://www.med.nyu.edu/
    content?ChunkIID=96921

Rosenthal, Elizabeth, "The Maligned Flatfoot: Some See an advantage." *The New
    York Times*, November 22, 1990

Rowlett, Ross, "How Many? A Dictionary of Units of Measurement." February
    23, 2001 & July 11, 2005. http://www.unc.edu/~rowlett/units/custom.html

Runyon, Damon, "The Tents of Trouble." Internet Archive. http://www.archive.
    org/stream/tentstrouble00runygoog/tentstrouble00runygoog_djvu.txt

Schmid, Randolf E., "Kin of Famous Lucy Had Feet Like Modern People."
    *USAToday.com*, February 11, 2011. http://www.usatoday.com/tech/
    science/2011-02-11-lucy-feet_N.htm

"Scholl, Now You're Walking." http://www.tobiasmayer.com/inf_scholl.php?IC=

"Sinus Tarsi Implant Insertion for Mobile Flatfoot," NICE Interventional
    Procedure Guidance, National Institute for Health and Clinical
    Excellence UK, issued July 2009. http://publications.nice.org.uk/
    sinus-tarsi-implant-insertion-for-mobile-flatfoot-ipg305/the-procedure

Slater, Lauren Slater, "Monkey Love." *Boston.com News*, March 21,
    2004. http://www.boston.com/news/science/articles/2004/03/21/
    monkey_love/?page=full

"Stephen King Trivia." StephenKing.com, http://www.stephenking.com/the_
    author.html

"The Three Spinning Women, Jacob and Wilhelm Grimm." Revised September 2,
    2002. http://www.pitt.edu/~dash/grimm014.html

Streicher, Julius, "How to Identify a Jew." *Der Giftpilz*. http://
    thecensureofdemocracy.150m.com/mushroom2.htm

"Talk: Flatfeet." Wikipedia modified May 13, 2012. http://en.wikipedia.org/wiki/
    Talk%3AFlat_feet

Tynes, Teri, "Gumshoes: A Partial Lineup of New York Detectives in American
    Crime Fiction." *Walking Off the Big Apple*, October 7, 2009. http://www.
    walkingoffthebigapple.com/2009/10/gumshoes-partial-lineup-of-new-york.
    html

"UNC Study: Frequency of Foot Disorders Differs between African Americans
    and Whites." *UNC School of Medicine,* November 8, 2010. http://www.med.
    unc.edu/www/newsarchive/2010/november/unc-study-frequency-of-foot-
    disorders-differs-between-african-americans-and-whites

"What Is Tarsal Coalition?" Seattle Children's Hospital. http://www.
    seattlechildrens.org/medical-conditions/bone-joint-muscle-conditions/
    feet-leg-malformations/tarsal-coalition/

Williams, Tara D., "Stranger Reaction in Children." Department of Psychology, Missouri Western State University, April 30, 2002. http://clearinghouse. missouriwestern.edu/manuscripts/341.php

Winter, John, "Footing the Bill for an Unusual Find." December 9, 2011. http://www.johnwinter.net/jw/2011/12/ footing-the-bill-the-story-of-an-unusual-find/

"A World History of Measurement and Metrics." http://www.cftech.com/ brainbank/otherreference/weightsandmeasures/MetricHistory.html

Yan, Cathy, "Are Asian Feet Different?" *FeminineBeauty.com*, January 11, 2011. http://www.femininebeauty.info/ethnic-comparisons/physique

"What Is Tarsal Coalition?" Seattle Children's Hospital. http://www. seattlechildrens.org/medical-conditions/bone-joint-muscle-conditions/ feet-leg-malformations/tarsal-coalition/

**4  •  Diet**

"Aretaeus of Cappadocia." *Encyclopædia Britannica*. Encyclopædia Britannica Online. June 29, 2012. http://www.britannica.com/EBchecked/topic/33531/ Aretaeus-Of-Cappadocia

L'Art de Soigner (advertismement), Bauman Rare Books, http://www. baumanrarebooks.com/rare-books/marie-antoinette-laforest-nicholas-laurent/l-art-de-soigner-les-pieds/83133.aspx

Ballentine, Carol, "Taste of Raspberries, Taste of Death, The 1937 Elixir Sulfanilamide Incident." *FDA Consumer*, June 1981. http://www.fda.gov/ AboutFDA/WhatWeDo/History/ProductRegulation/SulfanilamideDisaster/ default.htm

Battison, Leila, "British Flowers Are the of a New Cancer Drug." *BBC News Science and Environment*, September 12, 2011. http://www.bbc.co.uk/news/ science-environment-14855666

Bennett, James, "Cosmetics and Skin, Maybelline." 2012. http:// cosmeticsandskin.com/companies/maybelline.php

Binns, Corey, "Fact or Fiction?: No Big Toe, No Go," Scientific American, May 3, 2007. http://www.scientificamerican.com/article.cfm?id=no-big-toe-no-go

"Bradford Researchers Design Cancer Treatment to Safely Target All Solid Tumours." September 12, 2011. http://www.brad.ac.uk/mediacentre/press-releases/title-54781-en.php

Bragg, William Harris, "Noble W. Jones (ca. 1723–1805)." *The New Georgia Encyclopedia*, September 9, 2002. http://www.georgiaencyclopedia.org/nge/ Article.jsp?id=h-679

Brand, Richard A., "50 Years Ago in CORR: Osteoarthritis of the Hip in a Gorilla: Report of a Third Case, Robert M. Stecher MD CORR 1958," *Clinical Orthopaedics and Related Research*, January 2009, 467(1): 305–7. http://www. ncbi.nlm.nih.gov/pmc/articles/PMC2601004/

"Chaplin at Keystone: The Tramp Is Born." CharlieChaplin. com. http://www.charliechaplin.com/en/filming/ articles/212-Chaplin-at-Keystone-The-Tramp-is-Born

Choi, Hyon K. , Karen Atkinson, Elizabeth W. Karlson, Walter Willett, and Gary Curhan, "Alcohol Intake and Risk of Incident Gout in Men: A Prospective Study." *Lancet*, April 17, 2004, 363(9417). http://www.ncbi.nlm.nih.gov/ pubmed/15094272

Choi, Hyon K. , Karen Atkinson, Elizabeth W. Karlson, Walter Willett, and Gary Curhan, "Purine-Rich Foods, Dairy and Protein Intake, and the Risk of Gout in Men." *New England Journal of Medicine*, March 11, 2004, 350(11):1093–103. http://www.nejm.org/doi/full/10.1056/NEJMoa035700

Choi, Hyon K., and Gary Curhan, "Soft drinks, fructose consumption, and the risk of gout in men: prospective cohort study." *British Medical Journal*, February 9, 2008, 336:309–12. http://www.ncbi.nlm.nih.gov/ pubmed/18244959

Colchicine. Drugs.com. http://www.drugs.com/search. php?searchterm=colchicine

"Constitution of the United States, A History." http://www.archives.gov/exhibits/ charters/constitution_history.html

Cottrill, Jeff, "Sherlock Holmes Creator's 'Elementary' First Book to Be Published." *DigitalJournal*, June 7, 2011. http://digitaljournal.com/ article/307687#ixzz1itJNIL7U

Crisley, Kerry, "Fixing the One Dumb Thing That Benjamin Franklin Did." *Cool Green Science*, July 4th, 2011. http://blog.nature.org/2011/07/ fixing-the-one-dumb-thing-that-benjamin-franklin-did/

"CT Scans Can Help Detect Gout Cases Traditional Tests Miss." Paper presented at the *ACR/ARHP Annual Scientific Meeting*, November 5–9, 2011. http:// www.newswise.com/articles/view/582576/?sc=mwtr&xy=5003042

Dunn, Jimmy (John Warren), "Jean Francois Champollion, The Father of Egyptology." Tour Egypt. http://www.touregypt.net/featurestories/ champollion

Eamon, William, "The Tale of Monsieur Gout." *Bulletin of the History of Medicine*, 1981, 55: 564–7. http://williameamon.com/?p=414

"Editorials: Controversies in Family Medicine." *American Family Physician,* June 15, 2011, 83(12):1380–1390. http://www.aafp.org/afp/2011/0615/p1380.html

"Fascinating Facts About the Constitution," *ConstitutionFacts.com.* http://www. constitutionfacts.com/?section=constitution&page=fascinatingFacts.cfm

Fernie, W.T., *Meals Medicinal.* John Wright & Co., Bristol, England, 1905.

Finch, Jacqueline, "The Ancient Origins of Prosthetic Medicine." *The Lancet*, February 12, 2011, 377(9765): 548–49. http://www.thelancet.com/journals/ lancet/article/PIIS0140-6736(11)60190-6/fulltext

Fischer, Emil Hermann. Encyclopedia.com. Available online at http://www. encyclopedia.com/topic/Emil_Herman_Fischer.aspx

Frecklington, Mike, Keith Rome, Peter McNair, and Nicola Dalbethain, "Disability and Impairment Associated with Podiatric Problems in Patients with Acute Gout." *Supplement Proceedings of the Australasian Podiatry Council Conference 2011*. http://www.ncbi.nlm.nih.gov/pmc/articles/PMC3102996/

Gannon, Megan, "Ancient Fake Toes May Be World's Oldest Prosthesis, Study Shows," *LiveScience*, October 3, 2012. http://www.huffingtonpost.com/2012/10/03/ancient-fake-toes-oldest-prostheses-egypt_n_1936219.html

"Gout: Topic Overview." WebMD.com updated November 12, 2010. http://arthritis.webmd.com/tc/gout-topic-overview

"A Guide to 18th-Century Foot Care, Marie Antoinette's Copy, Beautifully Bound with Her Gilt Arms." http://www.baumanrarebooks.com/rare-books/marie-antoinette-laforest-nicholas-laurent/l-art-de-soigner-les-pieds/83133.aspx

Hanink, Johanna, "Orestis Karavas, Lucien et la Tragédie." *Bryn Mawr Classical Review*, 2006. Available on line at http://bmcr.brynmawr.edu/2006/2006-04-15.html

Hernberg, Sven "Lead Poisoning in a Historical Perspective." *American Journal of Industrial Medicine*, 2000, 38: 244–54. http://www.rachel.org/files/document/Lead_Poisoning_in_Historical_Perspective.pdf

Hippocrates, *Aphorisms, Section VI*, Trans. Francis Adams. http://classics.mit.edu/Hippocrates/aphorisms.6.vi.html

"A History of Gout (part 1). From the B.C. Centuries to the End of the 19th Century." Best-Gout-Remedies.com. Reviewed August 31, 2011. http://www.best-gout-remedies.com/historyofgout.html

"The History of HIV and AIDS in America." Avert. http://www.avert.org/aids-history-america.htm

"History of Wheelchairs." Chairdex. http://www.chairdex.com/history.htm

Johnson, Steven, "Green Ben: Benjamin Franklin and ecosystems." HistoryNet.com, June 4, 2009. http://www.historynet.com/green-ben.htm

Kesselheim, Aaron S., and Daniel H. Solomon, "Perspective: Incentives for Drug Development—The Curious Case of Colchicine." *New England Journal of Medicine*, April 14, 2010, 362:2045–47 (updated April 21, 2010 at NEJM.org). http://www.nejm.org/doi/full/10.1056/NEJMp1003126

Kippen, Cameron, "A Potted History of Podiatry." *Foot Talk*, December 22, 2008. http://foottalk.blogspot.com/2008/12/potted-history-of-podiatry.html

Kippen, Cameron, "Retifism and Fetishism: The Foot in Art." Blogspot, October 30, 2008. http://toeslayer-retifismandfetishism.blogspot.com/2008/10/foot-in-art.html

Kolata, Gina, "Study Discovers Road Map of DNA: A Key to Biology." *The New York Times*, September 6, 2012.

LaPook, Jonathan, "FDA Approval of Ancient Remedy Sends Price Soaring."

CBS News, October 10, 2011. http://www.cbsnews.com/2100-18563_162-20118283.html

"Left Atrium, Illness and Metaphor: Gout." CMAJ. January 23, 2001; 164(2): 239. http://www.cmaj.ca/content/164/2/239.2.full

Lewis, Jack, "Lead Poisoning: A Historical Perspective." *EPA Journal*, May 1985. http://www.epa.gov/history/topics/perspect/lead.html

Manno, Rebecca,"Gout." The Johns Hopkins Arthritis Center. http://www.hopkins-arthritis.org/arthritis-info/gout/

Matson, John, "Observations. What's the Real Story with Newton and the Apple? See for Yourself." *Scientific American*, January 18, 2010. http://ttp.royalsociety.org/accessible/SpreadDetails.aspx?BookID=1807da00-909a-4abf-b9c1-0279a08e4bf2

Maynard, Janet W., Mara A. McAdams, Alan N. Baer, Allan C. Gelber, and Josef Coresh, "Racial Differences in Gout Risk and Uric Acid Levels in Both Men and Women in the Atherosclerosis Risk in Communities (ARIC) Study." *Arthritis & Rheumatism*, 2010, 62(Suppl 10):1556. http://www.blackwellpublishing.com/acrmeeting/abstract.asp?MeetingID=774&id=90201

*Merck's 1899 Manual of the Materia Medica*. Merck & Co, New York, 1899.

de Montaigne, Michel, "Of Cripples," *Essays of Montaigne*, Vol. 9 [1580]. *The Online Library of Liberty*. http://oll.libertyfund.org/?option=com_staticxt&staticfile=show.php%3Ftitle=1750&chapter=91301&layout=html&Itemid=27

National Academy of Sciences, *Safety Testing*. The National Academies Press, Washington, DC, 2012. http://www.nap.edu/openbook.php?record_id=10733&page=21

The National Institute of Advanced Industrial Science and Technology (AIST), "Creation of 'Smallest Protein' Consisting of Only Ten Amino Acid Residues." August 10, 2004. http://www.aist.go.jp/aist_e/latest_research/2004/20040819/20040819.html

Nicholson, Geoff, "My Literary Malady," *The New York Times*, August 3, 2008. http://www.nytimes.com/2008/08/03/books/review/Nicholson-t.html?adadxnnl=1&adxnnlx=1321985624-CZTczNzAZR9/fvB5bNtWkw

Nuki, George and Peter A. Simkin, "A Concise History of Gout and Hyperuricemia and Their Treatment." *Arthritis Research & Therapy*, 2006, 8(Suppl 1):S1 . http://arthritis-research.com/content/8/S1/S1

Ombrello, T. "Plant of the Week: Autumn Crocus." http://faculty.ucc.edu/biology-ombrello/pow/autumn_crocus.htm

Omole, O.B., et al., "The Evolution of Gout (An Old Lifestyle Disease)." *South African Family Practice*, 2009, 51(5): 396–398 http://www.ajol.info/index.php/safp/article/viewFile/48488/34841

"Oops," *David Perdue's Charles Dickens Page*. Updated June 4, 2012. Pagehttp://charlesdickenspage.com/copperfield.html

"Opium." Medical Discoveries. http://www.discoveriesinmedicine.com/Ni-Ra/ Opium.html

Paddock, Catherine, "Gout Risk Linked To Genes." *Medical News Today*, March 10, 2008. http://www.medicalnewstoday.com/articles/100073.php

"Pseudogout." Mayo Clinic.com. http://www.mayoclinic.com/health/pseudogout/ DS00717

Reddy, Aravind, and Charles L. Braun, "Lead and the Romans." *Journal of Chemical Education*, 2010, 87(10):1052–5. http://pubs.acs.org/doi/ abs/10.1021/ed100631y

Rolleston, J.D., "Medical Aspects of the Greek Anthology." http://www.ncbi.nlm. nih.gov/pmc/articles/PMC2003554/pdf/procrsmed00948-0036.pdf

Roach, John, "Friday the 13th Phobia Rooted in Ancient History." *National Geographic News*. Updated August 12, 2004. http://news.nationalgeographic. com/news/2004/02/0212_040212_friday13.html

Safire, William, "On Language; Gimme the Ol' White Shoe." *The New York Times Magazine*, November 9, 1997. http://www.nytimes. com/1997/11/09/magazine/on-language-gimme-the-ol-white-shoe. html?pagewanted=all&src=pm

Scholtens, Martina, "The Glorification of Gout in 16th- to 18th-century literature,"Century Literature." *CMAJ,* October 7, 2008, 179(8). http://www. cmaj.ca/content/179/8/804.full

Schneider, Ben R., ed., "Seneca's *Epistles* Volume III, Epistle XCV," *Materials for the Construction of SHAKESPEARE'S MORALS: The Stoic Legacy to the Renaissance*. Revised January 29, 2001. http://www.stoics.com/seneca_ epistles_book_3.html#'XCV1

Schuler, Burton, "Dr Dudley J. Morton, The Father of the Morton's Toe." Foot Care for You. http://www.footcare4u.com/category/dr-dudley-morton/

Swiss Institute of Bioinformatics, "ProtParam for Human Titin." Expasy Proteomics Server. http://web.expasy.org/cgi-bin/protparam/ protparam1?Q8WZ42@1-34350@

University of Leicester, *Historical Directories*. http://www.historicaldirectories. org/hd/ud/usingdir4.asp

Wilkins, Robert H., "Neurosurgical Classic-XVII Edwin Smith Surgical Papyrus. *Journal of Neurosurgery*, March 1964, 240–44. http://www.neurosurgery.org/ cybermuseum/pre20th/epapyrus.html

Wilson, Jennifer Fisher, "Life Science. Got Gout? More and More People Do. Why?" The Smart Set from Drexel University, July 28, 2008. http://www. thesmartset.com/article/article07280801.aspx

Wood, George B., and Franklin Bache, *The Dispensatory of the United States of America,* 12th ed. J.B. Lippincott and Co., Philadelphia, PA, 1868

## 5 • Desire

"About us." UNOS.org, 2012. http://www.unos.org/about/index.php

Acar, Feridun, Sait Naderi, Mustafa Guvencer, Ugur Türe, and M. Nuri Arda, "Herophilus of Chalceaddon: A Pioneer of Neuroscience," *Neurosurgery,* April 2005, 56(4), 861–67. http://www.saitnaderi.com.tr/pdf/tarih/PDF46. pdf

American Psychiatric Association, "U 01 Fetishistic Disorder," *DSM-5 Development.* Updated April 29, 2012. http://www.dsm5.org/ proposedrevision/pages/proposedrevision.aspx?rid=63

Angier, Natalie, "Michelangelo, Renaissance Man of the Brain, Too?" *The New York Times,* October 10, 1990. http://www.nytimes.com/1990/10/10/arts/ michelangelo-renaissance-man-of-the-brain-too.html

Arnoux, Dominique J., "Coprophilia." *International Dictionary of Psychoanalysis.* The Gale Group, Farmington Hills, MI, 2005. http://www.answers.com/ topic/coprophilia

Ashliman, D.L., "Cinderella, Aarne-Thompson-Uther Folktale Type 510A and Related Stories of Persecuted Heroines." http://www.pitt.edu/~dash/ type0510a.html

"Bound Feet, the History of a Curious, Erotic Custom." http://www.josephrupp. com/bfindex2.html

Burton, Elizabeth C., and Kim A. Collins, "Religions and the Autopsy." *Medscape.* Updated March 21, 2012. http://emedicine.medscape.com/ article/1705993-overview

Baudelaire, Charles, "La Chevalure." *Fleurs du Mal/Flowers of Evil,* 1857 edition. http://fleursdumal.org/poem/203

Crites, James, A., "Foot Binding." *AngelFire,* October 25, 1995. http://www. angelfire.com/ca/beekeeper/foot.html

Debernardi, A., E. Sala, G. D'Aliberti, G. Talamonti, A.F. Franchin, and M. Collice, "Alcmaeon of Croton." Neurosurgery, February 2010, 66(2):247–52. http://www.ncbi.nlm.nih.gov/pubmed/advanced

"Did Michelangelo Draw A Brain in God's Neck?" *NPR,* June 21, 2010. http:// www.npr.org/templates/story/story.php?storyId=127990450

Eilberg-Schwartz, Howard, ed., *People of the Book.* State University of New York Press, Albany, New York, 1992.

"Evolution of Primate Sense of Smell and Full Trichromatic Color Vision," PLoS Biology, January 20, 2004, 2(1): e33.

Fields, R. Douglas, "Michelangelo's Secret Message in the Sistine Chapel: A Juxtaposition of God and the Human Brain." *Scientific American,* May 27, 2010. http://blogs.scientificamerican.com/guest-blog/2010/05/27/ michelangelos-secret-message-in-the-sistine-chapel-a-juxtaposition-of-god-and-the-human-brain/

"François Emile Matthes." http://www.yosemite.ca.us/library/matthes/francois_matthes.html

Freud, Sigmund. 1964. "Fetishism" (1927), in *The Standard Edition of the Complete Psychological Works of Sigmund Freud*, James Strachey with Anna Freud, eds. Hogarth Press, London, vol. 21: 152–57.

Furlow, F. Bryant, "The Smell of Love." *Psychology Today*, March 1, 1996. Last reviewed on August 13, 2010. http://www.psychologytoday.com/articles/200910/the-smell-love

Giannini, A.J., Colapietro, G., Slaby, A.E., Melemis, S.M., Bowman, R.K. (1998) "Sexualization of the Female Foot as a Response to Sexually Transmitted Epidemics: A Preliminary Study." *Psychological Reports*, 1998, 83(2):491–98.

"Guide to Understanding Islam." The Religion of Peace.com. http://www.thereligionofpeace.com/Quran/007-veils.htm

Halbach, Sara, Susan Pei, and Darrel Waggoner, "Caudal Appendage and Limb Abnormalities Are A Recurring Pair of Birth Defects." University of Chicago Genetic Services, March 12, 2012. http://dnatesting.uchicago.edu/blog/caudal-appendage-and-limb-abnormalities-are-recurring-pair-birth-defects

"Heikes Heels Page." http://www.heikes-heels.de/english/history-shoes/1.htm

"High Heels History Lesson." *Red Shoes for Women*. http://redshoesforwomen.com/red-shoes-articles/high-heels-history-lesson/

*The Holy Bible, King James Version*. American Bible Society, New York, 2009.

*The Holy Scriptures*. The Jewish Publication Society of America, Philadelphia, PA: 5712-1952.

"Is There a Mystery Hidden in the Sistine Chapel?" June 24, 2005. http://abcnews.go.com/GMA/Technology/story?id=878055

Jannini, Emmanuele A., e-mail correspondence. August 14, 2012.

"Jimmie Choo Heist." *Our Town*, August 30, 2012.

Kell, Christian A., Katharina von Kriefstein, Alexander Rosler, Andreas Kelineschmidt, and Helmut Laufsi, "The Sensory Cortical Representation of the Human Penis: Revisiting Somatotopy in the Male Homunculus." *The Journal of Neuroscience*, June 22, 2005, 25 (25), 5984–5987. http://www.jneurosci.org/content/25/25/5984.full

Lauren, Kay, "Heel appeal,'" *Dance Magazine*. http://www.dancemagazine.com/issues/December-2008/Heel-Appeal

Lim, Louisa, "Footbinding: From Status Symbol to Subjugation." *NPR*, March 19, 2007. http://www.npr.org/templates/story/story.php?storyId=8966942

Luering, H.L.E., "Foot." *Bible History Online*. http://www.bible-history.com/isbe/F/FOOT/

"Martha McLintock." Institute for Mind and Biology, University of Chicago. http://imb.uchicago.edu/people/members/mcclintock.shtml

Meshberger, Frank Lynn, "An Interpretation of Michelangelo's Creation of Adam Based on Neuroanatomy." *The Journal of the American Medical Association*,

October 10, 1990 264(14): 1837–1841. http://jama.jamanetwork.com/article.aspx?articleid=383532

Orr, James, M.A., D.D., Gen. Ed., "Definition for 'FOOT.'" *International Standard Bible Encyclopedia*. 1915. http://www.bible-history.com/isbe/F/FOOT/

Oztekin, H.H., H. Boya, M. Nalcakan, and O. Ozcan, "Second-toe length and forefoot disorders in ballet and folk dancers." *Journal of the American Podiatric Medical Association*, September–October, 2007, 97(5):385–88. http://www.ncbi.nlm.nih.gov/pubmed/17901343

Patterson, Richard D., "The Earth Is My Footstool, God's Feet and Our Walk." Bible.org. http://bible.org/seriespage/%E2%80%9C-earth-my-footstool%E2%80%9D-god%E2%80%99s-feet-and-our-walk

Randall, Gary C., and Janet A. Randall, "The Developing Field of Human Organ Transplantation," *Gonzaga Law Review*, Fall 1969, 5(20):20–39. http://gonzagalawreview.org/files/2011/11/gonlr5.7.pdf

"The Renaissance." History of Footwear. http://www.footwearhistory.com/highrenwomens.shtml

Saraf , S. and R.S. Parihar, "Sushruta: The First Plastic Surgeon in 600 B.C.." *The Internet Journal of Plastic Surgery*, 2007, 4(2). http://www.ispub.com/journal/the-internet-journal-of-plastic-surgery/volume-4-number-2/sushruta-the-first-plastic-surgeon-in-600-b-c.html

Scorolli, C., S. Ghirlanda, M. Enquist, S. Zattoni, and E. A. Jannini, "Relative Prevalence of Different Fetishes." *International Journal of Impotence Research,* 2007. 19:432–37. http://www.ncbi.nlm.nih.gov/pubmed/17304204 and http://www.nature.com/ijir/journal/v19/n4/fig_tab/3901547t2.html & personal correspondence

Seales, Rebecca, "Let Them Wear Heels! Marie Antoinette's Shoes Fetch £36,000 at Auction." *The Daily Mail*, March 29, 2012. http://www.dailymail.co.uk/news/article-2122012/Marie-Antoinettes-shoes-fetch-36-000-auction-Toulan.html#ixzz24mW4Cuqt

"Shoe History: The History of Your Shoes." ShoeInfoNet. www.shoeinfonet.com/aboutshoes/history/historyyourshoes/histor0yourshoes.htm

Spinney, Laura, "Five Things Humans No Longer Need." *New Scientist*, May 19, 2008. http://www.newscientist.com/article/dn13927-five-things-humans-no-longer-need.html

Steinberg, David, "Where is Your Heart? Some Body Part Metaphors and Euphemisms in Biblical Hebrew." http://www.adath-shalom.ca/body_metaphors_bib_hebrew.htm

Suk, Ian, and Rafael J. Tamargo "Concealed Neuroanatomy in Michelangelo's Separation of Light from Darkness in the Sistine Chapel." *Neurosurgery,* May 2010, 66(5):851–61. http://www.ncbi.nlm.nih.gov/pubmed/20404688

Tumolillo, M. Amedeo, and Robert Mackey, "The Lede: Saving a Notorious Shoe Collection in the Philippines." *The New York Times*,

October 22, 2009. http://thelede.blogs.nytimes.com/2009/10/22/
saving-a-notorious-shoe-collection-in-the-philippines/

"V.S. Ramachandran's Tales of The 'Tell-Tale Brain.'"
*NPR Books*, February 14, 2011. http://www.npr.
org/2011/02/14/133026897/v-s-ramachandrans-tales-of-the-tell-tale-brain

"Walk to the Guillotine." Feminet, June 12. 2012. http://feminet.wordpress.
com/2012/06/12/walk-to-the-guillotine/

Wedekind, Claus, Sina Escher, Matthuijs Van de Waal, Elisabeth Frei, "The Major
Histocompatibility Complex and Perfumers' Descriptions of Human Body
Odors." *Environmental Psychology*, 2007, 5(3):330–43. http://www.epjournal.
net/wp-content/uploads/EP05330343.pdf

Wedekind, Claus, and Dustin Penn, "MHC Genes, Body Odours and Odour
preferences," *Nephrology Dialysis Transplantation*, 2000, 15(9): 1269–71.
http://ndt.oxfordjournals.org/content/15/9/1269.full

Wells, C.A. Harwell, "The End of the Affair? Anti-Dueling Laws and Social
Norms in Antebellum America." *Vanderbilt Law Review*, 2001, 54(4): 1805–
1843. http://papers.ssrn.com/sol3/papers.cfm?abstract_id=1121921

"Dr. Wilder Penfield." Library and Archives Canada. http://www.
collectionscanada.gc.ca/physicians/030002-2400-e.html

Wyngard, Amy S., "Defining Obscenity, Inventing Pornography: The Limits of
Censorship in Rétif de la Bretonne." *Modern Language Quarterly*, March
2010, 71:1. http://depts.washington.edu/mlq/contents/contents.php?id=71.1

**Appendix**

"Africa the Birthplace of Human Language, Analysis Suggests."
*ScienceDaily.com*, April 15, 2011. http://www.sciencedaily.com/
releases/2011/04/110415165500.htm

# INDEX

Newton, Isaac, 109
Nigg, Benno M., 85–86
Nikolaevich, Alexei, 35
Noel, Roden Berkeley Wriothesley, 40
nose, 128, 130–31
nuchal crest, 14
nudity, 141–42, 152, *153*

O
Octavius (emperor), 89
*Of Human Bondage* (Maugham), 44
olfactory receptors, 128, 130–31
O'Malley, Sean, 136
O'Neal, Shaquille, 89
O'Neil, Dennis, 17
opposable thumbs, 1, *2*
orthopedic surgery, 39, 41–45, 84–85
orthotics, 82–84, *83*, 85–86

P
*pais* (child), 158
*Pali Canon*, 68–69
Pan (god), *70*, 70–71
Pare, Ambroise, 38, 41
*Parts of Animals* (Aristotle), 140
Pasteur, Louis, 42
Patterson, Richard D., 134
Paul (disciple), 29
Pelletier, Pierre Joseph, 117
pelvis, 23–24, 25
Penfield, Wilder, 147–48, *148*
perception, 18–19, 81, 82, 144. *See also*
    ideal body
Perrault, Charles, 126
*pes* (foot), 88, 89, 158
phantom limb syndrome, 149
pharmaceuticals, 118–22
pheromones, 127–28, 130–31
Philip II (king), 109
phobias, 58–60
phrenology, 71–72
Picon, Molly, 79
*Le Pied de Fanchette* (Rétif), 153
Pitt, William, 110–11
Plato, 48
*pod* (foot), 159–60

podagra, 95, 107, 159
podiatry, 115
*Podunk* (Zimmerman), 74
Pollio, Marcus Vitruvius, 7–8, 88–89, 92
polydactyly, 51
Ponseti, Ignacio, 44–45
posture, 13–19, 22–27, 60
primates
    apes, 13–14, 17, 19–26
    monkeys, 24, 25, 59
proportion, 7–8, 71–72. *See also*
    measurement
prosthetics, 93
Protagoras, 86
proteins, 97–98
psychology
    anxiety, 58–60
    body measurements and, 71–72
    of fetishes, 145–47, 151, 153–54
    of love, 58–59
    skull size and, 71
    social anxiety, 59–60
Ptolemy I (king), 140
purines, 98–100, *99*

Q
quadrupeds. *See* animals

R
race and ethnicity
    discrimination, 47–51, 68, 73–75
    eugenics, 47–51
    foot variations by, 72, 75–76, 82
    gout statistics by, 107
Race Betterment Foundation, 49
*The Rake at Rose Tavern* (Hogarth), 152, *153*
Ramachandran, Vilayanur, 147, 149–50
Reage, Pauline, 155–56
*The Red Shoes* (Andersen), 154–55
reflexology, 113
religion
    art and, 69, *69*, 132–33, *133*, 140–42

17/13